Praise for Jim Crumley's natu

"The best nature writer working [...]
DAVID CRAIG, *The Los Angeles Times*

"At his best he's peerless, in a class of his own among current outdoor writers." JIM PERRIN, *The Great Outdoors*

"Enthralling and often strident." *The Observer*

"Compulsively descriptive and infectious in its enthusiasm."
Scotland on Sunday

"The writing remains glowing and compelling."
The Countryman

"Weaves historical research, fact and narrative, fictional dialogue, philosophical reflection, lyrical prose, nature writing, memoir, apologia, and even some poetry."
ALISON MILLER, *Scottish Review of Books*

"Every well-chosen word is destined to find its way into our hearts and into our minds and into our imaginations."
IAN SMITH, *The Scots Magazine*

"With the pen of a poet and the eyes of a naturalist Jim Crumley reminds us of the simplicity of being. We are carried along in a sense of wonder." CAMERON MCNEISH, *The Great Outdoors*

"Jim Crumley, like all the best nature writers, combines his extensive knowledge with respect and awe for the natural world. He brings a soft-voiced poetry to his observations."
The Herald

"Well-written... elegant. Crumley speaks revealingly of 'theatre-in-the-wild'." *Times Literary Supplement*

"Jim Crumley has spent years observing eagles...and writes about them with passion and poetry." MANDY HAGGITH

Also by Jim Crumley

To Aileen with best wishes

THE
EAGLE'S
WAY

JIM CRUMLEY

Jim Crumley

Saraband ◉

Published by Saraband
Suite 202, 98 Woodlands Road
Glasgow, G3 6HB, Scotland
www.saraband.net

ISBN: 9781908643476
ebook: 9781908643483

Printed in the EU on sustainably sourced paper.

Editor: Craig Hillsley
Cover illustration: Joanna Lisowiec
Text layout: Laura Jones

1   3   5   7   9   10   8   6   4   2

CONTENTS

REMEMBER WELL

Andrew Currie

(1930 – 2012)

Over the hill and so far away
The eagle is soaring in full feather display
And she's certain and sure in the wide open sky
Of some brave direction that her heart can't deny

Dougie MacLean, 'All Who Wander',
from the album *Resolution* (Dunkeld Records, 2010).
Published by Lime Tree Arts and Music.

PROLOGUE:

WHAT NOW?

NINE IN THE MORNING and the sun has already gone from the crag. All day now the eyrie will be in shadow. But she is a pale eagle, and she brings to that gloomily overhung north-east-facing rock ledge the luminosity of a ghost. When she settles low in the cup with her one surviving chick, her head is a pale outcrop on the nest's steep-sided pile of timbers and greenery, like a cairn on a mountaintop. When she stands and walks to the edge, she is slow and clumsy like a sideways duck; and eerie. When she steps off, flaps twice, and lays her wings wide and still on the mountain air, she sails from shadow into sunlight and she becomes in a transforming moment what ornithology says she is – golden.

For a few seconds of level flight she presents her slimmest profile, the taut, unfurled scope of her wingspan's leading edge, seven or eight feet of it from wingtip to wingtip, all of it made more memorable by her light tawny pallor.

She begins to cross the glen, pauses in mid-flight to shake herself from stem to stern, scattering the night's dew so that it puts a glittering halo about her that catches the sun, then fades and falls away in shimmering, dimming droplets, and she resumes her easy, level, gliding flight. She is a hundred feet above a certain rock when she banks and looks down, side-headed, at the quiet, mountain-shaded shape that looks up at her, that has been looking at her for hours now. Its watcher's eyes see through raised glasses a single golden eagle eye, a glint of hooded amber as the sunlight laves her.

A wide, banking turn realigns her flight from east to south-west, and for a moment the sun lights up her under-wings, and there her plumage is almost blonde and almost pale gold, and the watcher on the ground shakes his head in frank admiration. There is wonder at work here.

She turns her back on him, and still without a wingbeat she re-crosses the glen, folding into a shallow glide that builds formidable speed. The only thing that can live with her in this mood is her own shadow, which hurtles ahead of her among rocks and trees, races across the river and starts to climb the glen's west flank. Now she is two hundred yards south of the eyrie buttress, and she stalls on the air and drops to a yard above a chaotic terrace of rock, birch scrub and mountain grasses, all of it broken by burns and waterfalls, and there her shadow waits for her. Together they begin to inch south down the glen, rippling over the contours, her shadow always a yard ahead, towing her south. Her airspeed is the nearest that nature will allow her above zero knots; it has slowed and slowed until all that is left now is to stall, but she does not stall, and nor does she pause, and nor does her shadow.

They travel half an unswerving mile together. Then, abruptly, they part. She has flung herself up the face of a shaded buttress and when she climbs beyond it she is a black cut-out bird against the blue-white western sky. She levels out and her wings beat a dozen times as she starts to circle, and as she does so she begins to climb. She climbs and climbs until, in the watcher's glasses, she is the size of a blown leaf in an autumn wind, and just when he thinks she is lost to him she eases out of what proves to be the topmost circuit, levels again, half closes her wings and flies east again, bridging the glen's tall chasm in seconds, powering over the skyline behind the watcher's back, so that he must first twist then stand and turn to follow her flight.

And now she is an absence.

And now the glen has emptied of its most vital life force. She has loaned the watcher five minutes of her time, his reward for an early rise, an hour's climb, and four hours of waiting at the base of the rock where she knows he often waits. She does not know why he waits there, for he provides neither threat nor food, nor performs any useful function that she can detect. He is there from time to time and he waits and watches and she is indifferent to that.

☺☺☺

With the eagle's disappearance, an old familiar doubt descends and hovers over me, a cold cloud of misgiving out of a clear sky. The doubt takes the form of a question: what now? I am earth-bound on the upper-floor of the glen, the eagle is God knows where. Whenever I seek her out, the long hours of her days are mostly far beyond my reach, or she is a huddled pale blur on the eyrie. And on the good days she lends me five minutes of her time. Now I will climb to the watershed, scour the square miles of what I know of her territory, and perhaps our paths will cross again for a few moments or a few minutes more. My idea of her life is gleaned from scraps she lets fall like discarded feathers. I gather them up gratefully, make what I can of their meaning. Yet still I think that she embodies some kind of key for me, the means of my understanding of her world as she sees it, and of her portion of my own territory as I see it.

This matters.

In the wolf-less Scotland I was born into, and where I have lived and worked all my life, this eagle is nature's ambassador, the catalyst that stirs wildness into its most primitive endeavours. I think that if I cannot pin down some sense of her place in nature's scheme of things (and in my country in my lifetime that scheme of things is deprived of all the prime movers and shakers of northern hemisphere wilderness, and especially deprived of the wolf), I fail myself as a writer and I fail the very landscape where I have set out my stall. I need more from her, but she is eagle and I am a fragment of her landscape, sometimes travelling between known haunts of her territory, like this rock, like a small rowan on the watershed, like that far skyline rock where once I watched her mate sit more or less motionless for four hours. I never rationalised what he was doing there all that time and I learned nothing from him beyond my own limitations as a watcher: he was still there when I left in the dusk (he may have been there all night, and I wish I knew). Hence the old familiar cloud of doubt. It is – it should be – part of the nature writer's condition.

This notion troubles me more since I wrote my wolf book, *The Last Wolf* (Birlinn, 2010), and I became utterly convinced that in any northern hemisphere land like this one, nature needs the wolf above all else in order to exercise the full extent of its powers. But right here, right now, the eagle is all the wolf nature has to work with. So, what now?

Nature's terrain is threefold: the land itself (and the native forest in all its complexity and diversity is its preferred state of landscape), the ocean, and the air. This land has been emptied of all the big mammal predators by its people, and its forests all but emptied of ancient trees, also by the people. The ocean slowly empties of its whales and the people's fingerprints are on that outcome, too. It may be that these processes have reached some kind of nadir, a rock bottom where a sluggish enlightenment has begun to stir in the gloomiest depths, for here and there and all across the face of the land, disciples of nature are planting new trees and caring for old ones and tree-loving species flock to the first small symptoms of recovery. Beavers, those unsurpassable architects of nature, are back. A few of them were planned by way of a government-approved trial in Argyll, many more mysteriously contrived from the ether on the other side of the country on Tayside; however they got here, they are beginning to make their presence felt. Whales have more friends among the people than at any time in the last two hundred years, and these have begun to find their voice.

And in the air, the golden eagle still makes waves, still makes a difference, still rules on nature's wolf-less behalf. The eagle has always had human allies. The native population is just about holding its own. But a change is happening in the eagle's world. Even as the beaver begins to thrive again for the first time in perhaps three hundred years, even as naturalists and nature writers take heart and begin to talk more hopefully about the admittedly distant prospect of bringing wolves home for the first time in two hundred years, the pale eagle of my early morning vigil is at the centre of the new change. That change is happening fast, faster than nature could ever hope to achieve by its own devices. Because we, the people, the culprits of so many outrages against nature for so long, are putting back the white-tailed eagle – the sea eagle – to share the golden eagle's sky. Nature senses an opportunity.

My question as the eagle crosses the morning skyline acquires a double edge: what now? I stamp cold feet into the hillside and head for the watershed.

PART ONE

THE
CHANGE

Chapter 1

AT THE TOMB
OF THE EAGLES

A MILE SOUTH OF JOHN O' GROATS, two whimbrels looking like curlews down on their luck flew across high fields low and flat and fast (40 mph – I measured them for a few moments with my car's speedometer as their flight paralleled the road). Then Orkney rose up apparently out of the land, then the land fell away and Orkney lay on a sea the blue of goddess eyes. In my mind, goddess eyes are the shade of Pentland Firth blue where Orkney lies on a May morning of sunshine and skylarks. So why, on such a day, would I be looking for a tomb?

Whenever I travel to Orkney now I do so with two people on my mind, both called George, both rooted in these islands, both dead to my great sorrow, so journeys to Orkney have acquired a memorial edge.

One is George Mackay Brown, the gracious Bard of this place, and in whose words Orkney travelled the world – still travels the world. Admirers of the man's quietist, wondering writing, who seek out Orkney in something of the nature of pilgrimage, have had their eyes opened on an Orkney not seen by others, a quietist, wondering Orkney whose stories are told in symbols – cornstalk, skylark, tinker, skull, fish, harp, star, eagle. George Mackay Brown is, by a considerable margin, the writer I admire most, the inspirational well where I drink deepest. In particular, I reach again and again for his poetry, and especially his posthumous collection, *Travellers* (John Murray, 2001). Orkney is a place to which people have travelled for at least five thousand years, but George Mackay Brown was a still centre, like island bedrock. He travelled almost nowhere at all, but wrote down the travellers and their stories. One of his novels, *The Golden Bird* (John Murray, 1986), is a reworking in an Orkney setting of an old and widespread folk tale about a baby snatched by an eagle from the edge of a harvest field. Despite the book's title, the eagle that inspired the story was almost certainly a sea eagle, for historically Orkney was a sea eagle haunt and unlike the mountain-thirled golden eagle, the sea eagle was never afraid to mix it with humankind.

Then there is his poem, 'Bird and Island':

A bird visited an island,
Lodged in a cliff,
A stone web of mathematics and music.

Bird whirled, built, brooded on
Three blue eggs.

A bird visited the island,
Sun by sun, aloof

From wild pig and dolphin and fossil
But woven into
The same green and blue.

The bird returned to the island,
Saw curves of boat and millstone,
Suffered fowler and rock-reft.

The bird, sun-summoned,
Turned slow above
The harp, the fire, the axe.

Bird and boy
Shared crust and crab.

Bird brooded
On a million breaking rock songs.
Bird visited, hesitant,
The island of wheels.

Bird entered
The heavy prisms of oil.

Flame now, bird, in your nest
Of broken numbers.

The other George is George Garson, the closest friend I ever had, and in whose travelling company I was introduced to his friend George Mackay Brown in Stromness. George Garson was an artist — mosaics and stained glass, in particular — and head of murals and stained glass at Glasgow School of Art until he was scunnered by "the suits who took over the art game". He took early retirement in his mid-fifties, after which he wrote a bit, but drew and painted every day for twenty-five years until the last days before his death a few weeks before his eightieth birthday. He was Edinburgh-born but fiercely, even aggressively thirled to a long Orcadian lineage. Orkney was in his blood, he was accustomed to travelling to Orkney from earliest childhood and throughout his life (something of Orkney travelled everywhere in him), and Orkney stone in all its forms coursed through his art as island blood fed his veins and sustained his heart. When he discovered the possibilities of slate as a mosaic material, he wedded it to his intimate knowledge of Orkney geology and Orkney architecture from Maes Howe to St Magnus Cathedral, and fashioned art from it all, art rooted in Orkney stone. All of the above is caught in a mosaic of grey and black slate and stone called "Black Sun of Winter", which is in the collection of the National Museum of Scotland in Edinburgh. Its standing stone motif is decorated by runes and dominated by a haloed sun, a black sun, the whole set in columns of tight-packed stones. The title of the piece is a line from a George Mackay Brown poem. I imagine that if you were to cut through the stone, it would say "Orkney" all the way through.

So these are the faces and voices I memorialised in my mind as I followed the flight of two whimbrels a mile out of John o' Groats and I looked up and saw Orkney on a sea the blue of goddess eyes. This was a view of Orkney George Garson never saw, for he had only ever crossed from Scotland to Orkney by way of Scrabster to Stromness (and long ago by steamer from Leith), but this way lies Gills Bay and the newest ferry that crosses not to the Stromness of George

Mackay Brown's small flat with its rocking chair and its coal fire, but to St Margaret's Hope on South Ronaldsay. There, on that blue day of May, the Tomb of the Eagles awaited me, five thousand years after it had been expertly fashioned from the raw stuff of the island by no one knows who.

⊙⊙⊙

"Here, hold them," she said, and I felt the press of eagle talons in the palm of my hand. This is not the kind of thing that happens to me every day. The first time had been – what – thirty years ago? I held a golden eagle on my thickly gloved wrist. It was a falconer's bird, and I knew the falconer a bit, not well, but well enough for him to show me the eagle (it had a name – Samson) and say, "Here, hold it." The eagle saw what was going on, understood, and stepped from his hand to mine. I was unnerved, not by fear but by my awareness that the golden eagle is, by instinct and learning, a tribe that shuns man and all his works, yet this one was being commanded to make physical contact with me. I knew it had been bred in captivity, although I had no idea how, or how legally the egg or the chick had been obtained. An eagle does not stop being an eagle just because it is captive bred. Nothing about the moment was comfortable. I was unnerved on the eagle's behalf.

I watched the feathered feet shuffle from the falconer onto my outstretched arm, watched the talons curve and bite for grip, the bird balanced perfectly on its new perch, looking round, a flash of yellow-gold eyes beneath a hooded brow that seemed to frown. I would frown too in those circumstances. I couldn't feel the talons through the leather gauntlet.

What I could feel was the power of a grasp that breaks bones for a living. That and the weight of the bird on the end of my arm. But I watched the talons, the dull black gleam of them curving out and down from the well-spaced, down-curves of yellow toes; I watched the grey gleam of the beak's terminal hook and the yellow cere that curved out beneath the eyes; I watched the pale tawny-gold nape when the bird turned its head away, that mysterious shade of feathering that illuminates in strong sunlight in such a way that my species christened the bird "golden".

The falconer moved away north not long after that, and I heard with a sickening lurch of heart and mind that Samson had suffered a grotesque death such as only a captive bird can suffer. He was attached to a perch in the garden when a swarm of bees settled on him. He panicked, and he was stung to death.

The second time was – what – twenty years ago? I was driving over a quiet hill road when a car travelling in the other direction hit a barn owl in flight. The car drove on, the barn owl crumpled. I stopped, picked up the owl, thought it might not be quite dead, put it on a carefully folded fleece jacket on the back seat, and set off for the house of a friend about five miles away, a friend who had nursed countless injured birds and animals back to good health.

Halfway there, the owl revived, and confirmed its revival by appearing suddenly on the headrest of the front passenger seat. Instinctively I put out my left arm. Why, God knows. The owl stepped onto my wrist just as the golden eagle had done, but this time my wrist was bare, and the owl talons sank in as deep as talons can sink. The pain is barely describable.

I had been driving at about 40 mph, taking it easy. I now had to stop the car one-handed while my other hand felt as if (my frantic guess) I had plunged it into a bath of acid. The car stopped. The road was mercifully empty. I leaned the owl towards the passenger door and opened it. I tried to urge the owl out through the space. It dug in a little deeper. I changed tactics, opened the driver's door and eased myself backwards out of the car and into the road. The owl came with me, still attached to my wrist. I walked to a roadside fence post. I held the owl next to the top of the post and at once it stepped up from my wrist onto the post. Oh joy, oh blessed relief, oh thank you God, or whoever. Little rivulets of blood dribbled from my owl-inflicted stigmata.

I retreated slowly, watched the owl from the car. I wanted to be sure it could fly. It flew. I drove home, mostly one-handed.

Twice in the last thirty years, then, I have had talons on my wrist. Now, in a visitor centre a mile away from the Tomb of the Eagles at Isbister on South Ronaldsay, a woman with a delicious Orkney voice put four sea eagle talons in my hand and said, "Here, hold them." They were yellowed and blunt with age, and it is their age that is the point. They are, give or take a century, five thousand years old, and they were found carefully arranged alongside human bones in what the outside world now knows as the Tomb of the Eagles.

These talons flew. That is the first truth that stares up at me from where they huddle together in the cradling cup that forms spontaneously in the curved palm of my left hand, so like eggs in a nest. They also grabbed fish, duck, gannet, cormorant, swan, curlew, goose from the sea or the air and choked the life from them, lambs stillborn or weakly alive from the land; they left their imprint in snow and estuary mud. It is not possible to feel their strangeness as intimately as this and be indifferent to the individual bird that wore them, five thousand years ago. That the talons still exist at all in the twenty-first century AD is a consequence of the nature of the relationship on this island between eagles and its Stone Age people: it astounds us today, for we have no comparable relationship with eagles at all, but that particular tribe that lingered on South Ronaldsay for at least eight hundred years either side of 3000 BC knew that sea eagles could – and would – perform a sacred service for them. So one more truth about those talons that lie snugly in my hand is this: in their day they were also accustomed to anchor themselves deep into dead human flesh while the biggest hook beak in all nature ripped it open and tore it up into manageable, edible pieces. Today, we have a euphemism for the process – "sky burial". It is still practised among some of the world's more isolated tribes, in Tibet for example, where the dead are laid out on raised platforms on sacred mountains beyond the reach of earthly scavengers so that their mortal flesh will be cleansed from the bones by birds, especially eagles.

What makes Isbister different is that the sundered and scattered human bones were gathered up and buried, and alongside them were the bones, skulls and talons of the sacred birds that performed the sacred service – mostly the white-tailed eagles. There was no intact skeleton, either human or eagle, among the remains. Rows of talons were arranged alongside human skulls and broken torsos. George Mackay Brown's mysterious last line suddenly hit me: here was the "nest of broken numbers".

☉☉☉

So here then, is the beginning of my story. As things stand, the story of change that has begun to impose itself on the Scottish eagles of the twenty-first century goes no further back than this. The situation that I have begun to try and unravel has been evolving for five thousand years that I know of. It took the white-tailed eagle from sacred totem in 3000 BC to oblivion in the early twentieth century to reintroduction in the late twentieth century to role model of twenty-first century conservation and green tourism.

⊙⊙⊙

"Here, hold them," she said, and she put the talons in my hand and my palm curved into a cup to protect their eminence, their sanctity, while I stared at their lifelessness and willed some unknown something to speak to me, or if not to speak, to reach out to me with some sense of their meaning. Whenever in my life I have been moved beyond mortal comprehension, the kind of unreasoning heightening of the senses that is achieved in some people by, say, a religious revelation, it has been because nature has touched me in a way that I was unprepared for, and in some great or small measure my life is changed from that moment. Here was one more such moment.

It may seem an extravagant claim. After all, I had done nothing more than accept an invitation to hold four very small pieces of a long dead bird. But here is why I think they startled me. I had not come to Isbister as a casual visitor. I was not on holiday. I was not a tourist on a whistle-stop tourist trail. I had driven more than 250 miles to catch a ferry to South Ronaldsay to find the Tomb of the Eagles.

I did not know quite what to expect. I had read a little bit about it, and I cannot say that I have a great relish for tombs, or anything else that shuts out the overworld. I was propelled by a sense that the story of the book I wanted to write might prosper here. My friend George Garson had stepped this way one winter's day twenty years ago as part of his research for a very personal wee book called *Orkney All the Way Through* (John Donald, 1992), which saw Orkney with the eye of an artist and the sensibility of a Garson (a name that in its Norse-Orcadian inheritance meant "the son of the dyke-end", an inheritance he wore with some pride). It is fair to say that it was not the highlight of the book, for he was no one's idea of a naturalist or a nature writer, but he caught something of the lie of the land back then, before the visitor centre was built. His guide was Morgan Simison, the wife of farmer Ronnie, who had discovered what was to become known as the Tomb of the Eagles and who had done much of the early archaeological excavation himself. It was not eagle talons George was handed (for he and Morgan had obviously exchanged brief autobiographies). It was a human skull. He would write it down thus:

"Go on. Hold your grandfather's skull," she said, lifting the old boy from the glass case. "Now try this stone hammer." I cradled my grandfather in the crook of my left arm and grasped the smooth stone. It was snug and rightly made. "Beautifully balanced, eh? Now try this stone hammer. Fits the hand well, doesn't it? Your grandfather wasn't stupid, you know."

The tomb... rests on the cliff's edge between Black Geo and North Taing, its claustrophobic entrance facing seawards. The trick is to arrive there on your own, on a day of storms, and

*gulping fetid air, crawl into the past, time present snapping at the
soles of your feet.*

*Hugging the experience, I returned to the farm by the scenic
cliff path, secure in the knowledge that sea eagles breed – albeit
tenuously – along Scotland's western seaboard.*

That's it. That's all there is. Just a handful of whistle-stop
lines. Whatever he encountered at the far end of that solitary midwinter mile between farm and tomb, it unnerved
him, it was unnervingly more than he bargained for, and far
more than he was willing to write down. Instead, he stashed
it away within the chambered walls of a three-word phrase:
he left "hugging the experience". Nowhere else in his book,
and never once in hundreds of subsequent conversations and
dozens of letters in the thirty years I knew him, did he con-
ceal. He was an artist, and even when he wrote he painted
pictures. His instinct was to *reveal*. No man I ever met was
genetically less qualified to "hug" an experience. Yet when I
first read his guarded account of his visit to the Tomb of the
Eagles twenty years ago, none of that registered. It was his
first book (and his only one) and he was wildly enthused by
the process – by his research trips to Orkney, by discovering
a writing style, by the way *his* Orkney took shape before his
eyes and insinuated itself into what he wrote down. Our conversations, usually in a favourite Edinburgh pub, were awash
with Orkney for months – the places, the people, the sounds,
the travels, the stirred memories. But now that I think about
it, the Tomb of the Eagles did not surface once. It was only
when I set out on my own book's journey that I re-read
his take on the tomb, and instead of some anticipated shred
of enlightenment I found a shroud, and I wondered what

manner of truth he unearthed that day that he would never talk about it and never write it down. And of course it is too late to ask him now. But there is also this: now that I have been to the Tomb of the Eagles myself, I think I know.

◉◉◉

It was neither winter nor was it a day of storms when I followed in his footsteps those twenty years later, and two years after his death. Instead, it was May and sunlit and anthemed by skylarks and adagio curlews, and every footfall was dusted by ground-hugging, wind-cheating flowers – eyebright, primrose, spring squill, marsh orchid, grass of Parnassus. It is a mile to the tomb from the new visitor centre by the farm, with its sea eagle weathervane. I could see the tomb's inconspicuous shape in the distance but I knew it for what it was only because it was pointed out to me by a member of staff. In truth, it looked like nothing at all. Orkney on such a day is a many-layered wonder for a nature writer, and my mind was anywhere but inside a tomb. I walked from one oystercatcher territory to another, so the decibel levels of strident variations on a theme of "piss off" rose and fell about my ears with the rhythm of waves on shingle. Lapwings lined up strafing runs out of the sun and aimed them a few feet above my head. They also fell out with the nearest oystercatchers. Once, an Arctic skua, a darkly, beautifully lethal missile of a bird, sped above the fields trailing a raucous wake of six lapwings. Three oystercatchers joined in, united for the moment in common cause against the greater enemy. The sea shone, gannets fell and rose and dazzled, seals mourned. Above it all, skylark after skylark after skylark sang and sang and sang. By the time I got there I was drunk on a cocktail of island seductions and warm honeyed winds, and I was in no mood for tombs. Perhaps George was right: perhaps the trick is to go on a head-down day of midwinter storms. I looked at the black oblong of the knee-high entrance and inwardly I shuddered. What was it you wrote, George: "...*and gulping fetid air, crawl into the past, time present snapping at the soles of your feet...*"

You can crawl, or there is a weird, flat, wheeled trolley with a rope arrangement to ease your passage along the entrance tunnel. The tunnel is three metres long by eighty-five centimetres high by seventy centimetres wide. It is not a place to linger. I went in by trolley, head-first on my belly, which seemed the speedier option. The trolley deposits its cargo in the main chamber. I was at once grateful for two late-twentieth century skylights that were let into the late-twentieth century protective roof, but the place is still gloomily convincing enough for a tomb. Sunlight, sea and skylarks are elsewhere, in the overworld, lost beyond recall. You are supposed to be dead and in pieces to get in here. I stood up, dusted myself down, and looked round.

I don't like being underground. I don't like dark, enclosed spaces. I go out of my way to avoid them. Yet I had travelled more than 250 miles to reach this eerie here-and-now, and abandoned the skylark-scrolled, sun-smitten, ocean-scented island for a glimpse of the Stone Age from the inside of its corpse. All was grey stone, expertly stacked and fashioned into walls, alcoves, stalls, shelves; a mysteriously ordered silence wonderfully wrought in grey stone. And everything about it was at once utterly familiar. Every facet of the techniques at work, every trait and nuance, were as familiar to me as the walls of my own living room, because in fact they adorn the walls of my own living room in the form of two small slate mosaics by George Garson. And on the walls of George's house and in various exhibitions of his work down the years I have seen and studied their like again and again. In the Tomb of the Eagles I was astounded into something like a state of trance. I walked round the walls stopping after every yard, fingering worked stone from flat pebble-sized pieces to lintels a yard wide and slabs deployed as structural posts the size of tall men. When Ronnie Simison first forced a way in here, it was a mass grave of men, women, children and eagles. When I wheeled my way in it was tidied, bone-free, roofed, lit, and presentable to a tourist audience. But when George Garson "crawled into the past", he came face to face with something like his own ancestral ghost. When he stood up, dusted himself down and looked round, he would have seen himself, for the walls of the Tomb of the Eagles resemble nothing so much as a huge, walk-in George Garson mosaic. When he fingered these meticulously ordered stones, he was shaking hands not just with his ancestors but with the very origins of his own art, and his art was his life. That glimpse of his own mortality was why he had left hugging the experi-ence, "secure in the knowledge that sea eagles are breeding – albeit tenuously – along Scotland's western seaboard".

He would, perhaps, have been impressed at the nature of the bond between his ancestors and the ancestors of those tenuous eagles, for he loved simple, powerful symbolism. But it never drew him deeper into the eagle's world, and whatever insight he learned about Neolithic tribesmen and himself he kept it between himself and those who had gone before. I thought about him cradling that skull in the crook of his arm. Then his head-down, solitary tramp out to the headland where he crawled into the tomb and was tapped on the shoulder by five thousand years of his own lineage. Then I wondered what I was doing here. Then I stopped thinking for a while and just stood uneasily breathing the place in.

The next conscious thought I had was that I could still feel the talons nestled in my curved palm. They were not there, of course, for I had handed them back after a minute or so, then talked to several other people, then stepped from the visitor centre out into the sunlight, then walked a mile by field and clifftop, squill and skylark. But standing in that place of death, the life that these talons had led was a sudden, electrifying summons: *the relationship between man and eagle is reborn, you too are part of this.*

And just as suddenly the tomb's aura was too intense, too overloaded with too much past, and too enriched by my inadvertent collision with George's moment of revelation, and suddenly all that began to unnerve me, and I felt overwhelmed. I lay flat out on the trolley and sped head-first for the bright end of the tunnel like a breaching whale. Sunlight rushed in to meet me like helping hands.

I stood, dusted myself down, and without looking back, walked down to the shore, where the rocks are piles of flat tiers, and tilted like a half-open fan, and that still felt a little too much like the walls of the tomb (you can see how Orcadian geology informed the architects) so I climbed down to the ocean edge and watched a sea the blue of goddess eyes, where gannets and Arctic terns dive-bombed the shoals, and skuas cruised the wavetops, handsome thugs looking for unwary birds to mug.

It was half an hour before I turned to look inland again to where the tomb lay, just a low, green swelling on the surface of the island, apparently demure and sun-drowsed, and placid as a tidal pool at slack water. There must have been many eagles when the tomb-builders of the Eagle Tribe made landfall here, bearing their newly dead and their eagle-empathy. A good eagle population must have been their first consideration. I wonder if they came on a midwinter day of storms like my friend, or on a May day like this one of goddess eyes and skylarks. I walked back up to the tomb, admired how the restorers and nature had conspired to heal the walls and clothe the new roof in the green of the island, so that the tiered walls only showed here and there in patches. I was standing about a dozen yards from the entrance when a skylark rose from the very crown of the tomb's curved roof. Skylarks are my good omen birds. Where there are skylarks, there is hope. As long as there are skylarks, I can handle most things. And now a skylark sang at the Tomb of the Eagles.

The skylark that sang
at the Tomb of the Eagles

chiselled upwards a thin column
of runes, primitive truths

bound up in catchy slurs
and jazzy triplets, like Bechet

exploring the deep blues.
So it was when tomb-builders

made landfall at Isbister,
found biddable stone to chisel

runes and truths of their own,
and set aside a portion of headland

and the next eight hundred years
to memorialise the passage of their days

across the face of the island
domed in an unlettered grave.

The eagle's shrill anthem
was the struck harp of their song,

and talon-and-bone they honoured them
as they honoured their own.

Aarkum the Bard squinted skywards
under his raised hand

towards the rising improvising lark
and mouthed two prescient syllables:

"Besh-ay",
Song Island.

Skylark, eagle, builder, tomb —
it is all the same song.

It is all the same
unfinished song.

I walked back the spring-laden, cliff-girt mile to the visitor centre, walked through larksong after larksong after larksong, the song passed from singer to singer to singer. At the visitor centre shop I bought archaeologist John W Hedges's book *Tomb of the Eagles — A Window on Stone Age Tribal Britain*, (John Murray, 1987), in which he argues that a phenomenal amount of effort would have been required to build the structure, which consumed almost a thousand cubic metres of flagstone and other material. "It must have been a magnificent sight", he observes, and "it takes little imagination to appreciate the significance that it must have had for the generations that laboured over its erection, let alone those that followed. To know of one such monument belonging to so remote a period of antiquity is remarkable enough, but the fact is that no less than seventy-six are known in Orkney as a whole — of which Isbister must be accounted among the most splendid."

◎◎◎

That feeling in my left palm would recur unpredictably for months, and each time I would hurtle back to Isbister in my head, to the bright headland and to the cool, grey, flat-stone innards of the tomb for which the word "splendid" was not an option that came readily to mind. But the word that did come increasingly to my mind was "Alaska". Why Alaska? Because it was there in 1998 that I tried to interview a native Tlingit called John for a BBC Radio 4 programme I was making on the subject of the relationship between people and nature. He was a reluctant interviewee, which was understandable. After all, I had just stepped off a transatlantic plane, put a microphone in front of him and more or less asked him to bare his tribal soul. After a very protracted silence and several attempts to re-frame my first question, he finally spoke these words:

"It was not God who made the world, but Raven. Raven's first job was to make Nature in perfect balance so he made the Bald Eagle, with a white head, a black body, and a white tail to symbolise Nature in perfect balance..."

So to this day, the Tlingit and other native tribes of south-east Alaska all derive from one of two Clans – Raven and Eagle. Many aspects of ancient belief still colour their lives in the twenty-first century, alongside hefty pick-up trucks, laptops and mobile phones. They marry outwith their own Clan. Each new child takes their mother's ancestry. When an Eagle dies, a Raven watches over the body, and vice versa. But not everyone agrees with John's ordering of events. There is a certain amount of rivalry about which dynasty came first – Raven or Eagle. In the summer of 2012 that rivalry took a particularly contemporary twist at the biennial Celebration gathering of the tribes of south-east Alaska in Juneau, the state capital. Everyone attending the gathering was invited to participate in a DNA study to reveal something of the history of migration of people into Alaska. One consequence might actually determine whether or not Raven really did make the world and arrange for Nature's symbol of perfect balance, or whether the Eagle was already perfectly balanced when Raven got here. But the symbolism affected me deeply. Yet when I went to Alaska I knew almost nothing about the Tomb of the Eagles in Orkney beyond the fact of its existence and George Garson's perfunctory account, and nothing at all about the sacred relationship between man and eagle that it embodies. Now I can embrace both lands. Now I can begin to understand the poleaxing effect of four sea eagle talons in the palm of my hand.

George had been given a skull because he had told Morgan Simison about his grandfather and the nature of his visit. Another couple who were in the visitor centre museum at the same as I was, and who were simply on holiday and enjoying Orkney's historical sites were also given a skull to hold. I, who had said nothing to anyone about what I was doing there, might just as easily have been given a skull. Instead, I was given four sea eagle talons to hold. A life like mine turns on such moments.

The following month – June, 2012 – two emails arrived within a day of each other. One, from my friend and Alaskan nature writer of distinction, Nancy Lord, had attached to it an article from the *Anchorage Daily News* about the DNA of Ravens and Eagles. The other was an extract from a journal called *Bird Study* about the relative populations of golden eagles and sea eagles in Great Britain and Ireland. The thrust of the study was that persecution had been going on since around 500 AD, but what intrigued me more was that it included figures from 3000 BC, which is when the Isbister tomb was in operation and eagles were abundant enough to make possible the relationship between tomb builder and eagles. The numbers in the table here indicate breeding pairs.

DATE	GOLDEN EAGLE	SEA EAGLE
3000 BC	650	2550
500AD	1000–1500	800–1400
1800AD	300–500	150
1920	100–200	Nil
1950	280	Nil
1971	400	Nil
2003	440	31

Reintroduction of the sea eagle began on Rum in 1975. The final reintroduction programme began on the Tay estuary in 2007 and finished in 2012. The golden eagle population is more or less unchanged since 2003 and the sea eagle population has more or less doubled. It is, then, a matter of time before the sea eagle begins to outnumber the golden eagle again. We are never going to reclaim the kind of relationship that underpinned the Eagle Tribe of Isbister, but sooner rather than later we must come to terms once again with the idea of living with eagles in our midst.

Most of us have never seen a golden eagle. Those of us who know the bird know it for a haunter of the high and lonely places. The sea eagle can do that too, but it has no hang-up at all about perching on your roof. The golden eagle will have to learn to live with it too, but unlike ourselves, the golden eagle will not have forgotten how. The change has begun and its effects will become more and more clear from now on. The process is happening as I write. The situation will be different from what it is now by the time I finish writing this book.

Chapter 2

THE GRAPEVINE

THE WORD TRAVELS the grapevine in a life like mine. You establish your place in it slowly, over years. You feed in what you know, it feeds you in return. It is a careful, watchful process. If the word reaches the wrong ears, the consequences can be disastrous. Eagles might die. A whole landscape might fail. How you use the grapevine and its information is your business. If you misuse it, expect the grapevine to bypass you in the future and forever. Expect that at the very least.

Sometimes the information it offers you is inconsequential, clouds across the moon, but very occasionally it stops you in your tracks, and having stopped, you sense the intimation of a new thing beyond which all is changed. And I have heard about three young eagles roosting together in the hills above Loch Tay, and I have stopped in my tracks to consider the implications. And what I think is this: a new thoroughfare of nature stretching clear across the waist of the country has begun to build, and it is being built by eagles. It is the result of the decision to follow the white-tailed eagle reintroduction project on Rum (1975–85 and 1993–98) with a new phase (2007–2012) near the Tay estuary on the east coast, and the birds' various responses to their new surroundings.

So one day I heard about the three young eagles roosting in the hills above Loch Tay and I had questions for the grapevine.

"Eagles? Plural?"

I have watched many golden eagles in Highland Scotland over more than thirty years now. Young golden eagles travel widely especially in their first year, but they wander alone. Even siblings from the same eyrie leave the parent birds' territory separately and travel separately. As a rule they do not roost communally. They do not pluralise. At least not here. In North America they embark on long north-south migrations and flocks of golden eagles do occur. I can hardly imagine what that must look like. But three eagles in the same stand of trees on a hill above Loch Tay had to be something... other. The grapevine answered:

"Not golden eagles. Sea eagles."

"Sea eagles? Loch Tay?"

Loch Tay is heartland, the centre of the centre. There are few more utterly land-locked large lochs grasped in a fastness of mountains anywhere in the land. If you were to draw a line on a map of Scotland across its 130-mile-wide girth from the eastmost thrust of the Tay estuary to Mull in the west, it would bisect Loch Tay at the furthest distance from the sea in either direction; so on the face of it, a strange place for sea eagles.

☉☉☉

The morning was early, quiet and still. A half-hearted track took the steepening hillside head-on, no bends to ease the gradient. I am a slow starter on the hill. I like the chance to lubricate away the muscle-rust on easier slopes, I grow stronger with the morning. Also, steep ground encourages a head-down approach and mine is essentially a head-up, horizon-scanning gait. So I put images in my head of sea eagles I have known so that my eyes could watch the path. *Remember the first time?*

⊕⊕⊛

Mingulay, the Outer Hebrides, a few links in that island chain south of Barra, a warm late-May mist burned away over five days into a swimming-in-the-ocean early June heatwave, and I had gone for a long walk round Mingulay's hills. I considered an odd-looking rock two hundred yards away. It was as simple as that. One rock among thousands of rocks caught my eye because there was something not right about it. The nature writer behind that eye wanted to know why.

The rock had an embellishment on top, no less grey and unprepossessing than the rock itself, but it was somehow grey in a way that the rock was not, a different texture of grey, a ruffled grey, and browner than the grey of the rock, almost a pale tawny grey. It was as shapeless as the rock itself, but the surface of its shapelessness stirred a little in the breeze in a way that the surface of rock should not stir. It was also awkwardly poised on a top edge of the rock in a way that suggested it might topple at any moment. The rock was squat and rounded, about six feet high and at least as wide. The embellishment was about three feet high and half as wide. At first glance it looked perched and hunched, but at second glance I was thinking it was just another piece of rock and I had been deluded by a trick of the light. So I put down my pack, sat on it, raised the binoculars, and what slid into focus was the turned back of a huge, grey, perched, headless bird. I had simply never seen such a thing, never seen anything so distressingly headless, never seen anything so massively dishevelled (it had chaotically protruding feathers the size of paddles). I still thought it might topple over.

What changed everything was the screaming arrival of a pair of Arctic terns. Their low, fast, line-astern attack was aimed at where the headless bird's head should have been, and was impressive enough to induce it to duck, but when the creature straightened again it was quite a bit taller and it had a considerable neck, and a head. Then it turned to follow the flight of the terns and revealed an embellishment of its own, a hooked beak like a machete. I had seen many golden eagles before, including one unusually large female at unusually close quarters when we surprised each other, and that male that stood briefly on the end of my outstretched, heavily gloved arm; but this, I told myself, was not one of these. There was only one thing it could be.

I knew a fair bit about the sea eagle reintroduction project based on Rum, but I had never crossed the path of any of the fruits of its labours. So although I had read the book and seen the film and admired the photographs, none of that had readied me for the sheer monstrousness of this creature in my binoculars.

Twice more the terns attacked, twice more the eagle avoided a direct hit by dipping its head. But such was the disparity in the size of the birds and of the weapons at their disposal that I wondered frivolously if the terns were on a suicide mission. Then the eagle flew. It flew by deploying something like a half-opened parachute that immediately cloaked the entire rock in deep shadow, such an apparently uncoordinated acreage of wingspan that it looked as if chaos must result: the thing *would* topple over. Yet in the next moment it was clear the colossus was airborne and moving low across the hillside to a graceless, wing-heaving rhythm. A golden eagle flies from such a rock by stepping off and holding its wings wide and still, finding lift and grace at once. It is, I had assumed until that moment, how all eagles behave; it is, I was convinced before that moment, what makes eagles so admirable, the sheer control of every aspect of flight, so that nothing at all is beyond them, and no manoeuvre is conducted with anything other than fluency and style and, yes, grace. And then there was this... this *upheaval* of a bird, this massive apology for an eagle with all the grace of an airborne tank. But then I noticed something familiar, something I knew from all those years of watching golden eagles, something wonderfully exclusive to the tribe of eagles. As the sea eagle flew low above the surface of the land it was pursued across the island by the ominous black sprawl of its shadow that rippled over rocks, all those other rocks that simply amounted to the lie of the land, until my nature writer's eye had stumbled over the one that didn't look right. Watching the eagle devour distance, I saw how every rock it passed was darkly transformed as the touch of that improbable shadow crossed it and moved on. I thought: this is an eagle alright. Then I thought: and this is a bird that can make a difference. That was the first time, the first of all my sea eagles.

⊚⊚⊚

I paused for breath above Loch Tay, looked up and saw distant trees. Up there? Head down again. *Remember the second time?*

◎◎◎

The Isle of Mull, autumn, three years after Mingulay. I considered an odd-looking tree two hundred yards away. It was as simple as that. One tree among hundreds caught my eye because there was something not right about it. It shifted uneasily in a big wind. I was driving slowly on one of Mull's coast-hugging single-track roads. I had not seen a vehicle for some time and I was driving with a nature writer's eyes rather than a motorist's, watching for otters mostly (this because a few years earlier driving this same stretch of road, an otter had stepped from the ditch and trotted up the road ahead of the car and I had followed it in first gear for about a hundred yards before it took to the rocks of the shore). Then the road twisted among trees and there was that birch that shifted uneasily, then it spilled an eagle from the innards of its crown.

There was the same collapsing parachute aspect to the unfurling curve of wings that I had seen on Mingulay, the curious sense of a bird that looked too large for the details of the landscape. It drifted past the last tree and headed for the shore in a long glide.

The tide was far out, which on that particular corner of Mull means a wide expanse of mud was laid bare. The eagle landed, and went walking out across the mud. I wasn't clear why, as it didn't seem to be trying to feed there. When it stopped with its back turned and its giveaway beak out of sight, and with its raggedly disordered plumage and dark grey-brown colouring, it looked like a broken fence post. No golden eagle ever looked like that. There again, golden eagles don't generally go walking in tidal mud, even Mull golden eagles. I was learning quickly that the sea eagle is something of a shape-shifter, and that it is also a lot less wary of the human race and all its artefacts than the golden eagle. After a few minutes of mud-wading the bird flew again, and again I thought that as with the Mingulay bird, the flight at low level looked heavy, disappointingly graceless. No golden eagle ever flew like that.

It had travelled no more than fifty yards when it dislodged a second eagle from the mud, one that I had simply not seen because it too had assumed the guise of a broken fence post in the mud until it revealed itself by flying. They flew off together, low and slow, heading for the distant sea. I crossed the bay to the muddy terrain where I had seen the first eagle land, and there I saw my first eagle footprints. There must be occasions when golden eagles leave footprints in shallow snow, although I have never come across them in all my years of walking golden eagle country, years of watching golden eagles. But it seemed to me that day that sea eagle footprints were about to become a commonplace feature of tidal landscapes. They are, of course, huge.

☉☉☉

So those were my first two encounters with the sea eagle, more properly known as the European white-tailed eagle, *Haliaeetus albicilla* if you must, also known in antiquity as the erne. But "sea eagle" is the name that has stuck, although the bird is a prodigious traveller and can turn up almost anywhere there are tracts of open water. Like Loch Tay, for instance.

☉☉☉

The hills around Loch Tay, Highland Perthshire, in the spring. I paused a few hundred yards away from a group of old Scots pines, not tall but broad-shouldered and stout-limbed. I was responding to an invitation from Polly Pullar, writer, photographer and friend, to come and look at these trees in the early morning, because they were being used as a roost by sea eagles. I leaned against a shadowed rock and focussed binoculars on them. Early sunlight cross-hatched the limbs and branches with deep black shadows, making a patchy chaos of the canopies. No roosting eagle could wish for better daylight camouflage. I let the glasses roam slowly, let my eyes accustom to the shapes and shadows within the trees, looked for the shapes and shadows that weren't quite right, found none. I waited, I worked the glasses among the branches. It was still early. They could still be there, in which case, sooner or later they would fly, in which case they would stir, stretch their wings, shake the moisture from their plumage and reorder it, any or all of these things before they flew; or, as it happened, none of these things.

They came out, one by one, and minutes apart, without fuss or fanfare or any warning that I detected, and of course, they came out of the wrong side of the trees, the shaded side, the side away from the sun. Now why did I not think of that? But they drifted on the air and did not go far, and one perched on the hillside and preened in the sunlight, and two that had left five minutes apart and vanished west came back along the skyline wingtip to wingtip, and there were not three eagles, there were four. The fourth flew alone, and it was paler and smaller than the others and there was considerably less of the machete about its beak, and it flew more gracefully and with far fewer wingbeats, and the white under its wings and its tail confirmed the presence among the roosting young sea eagles of a roosting young golden eagle.

The thing about this part of the country is that it is halfway between the established sea eagle breeding territory of Mull and the relatively new east coast reintroduction project on the Tay estuary. All three places are on the same line of latitude. In the hills immediately north and south of the loch, and all the way to the west coast, the terrain is long-established golden eagle country. Sea eagles have been absent here for more than a hundred years. *Had* been absent...

The thought struck: pilgrims for nature, travelling the same line of latitude across the country but in opposite directions. They had all begun with no destination in mind. Another thought: was this the destination nature had in mind for them as they pursued some flightpath of inherited knowledge? In the new sea eagle strongholds of Mull and Skye, they frequently come into contact with golden eagles, sometimes hostile contact, but I had heard of nothing like this before.

All four roosting eagles were young birds, probably even first year birds. The golden eagle would lose the white underwing and tail patches in time and the sea eagles would gain the diagnostic white tail of their tribe. The sea eagles had come on this high roost above Loch Tay from the east where they had been released, and the golden eagle had most likely come from an established territory in the west from which it had been driven by its parents. *The pale eaglet of a pale mother,* I thought to myself. *The territory of the pale eagle I know so well is a leisurely hour's flight from here. Could this be...?*

The young wanderers had found common cause in each other's company; that was the next thought that struck. They wandered off about the day's business. There was no sign of any collective purpose but I do think they spent it travelling within sight of each other for much of the time. They had been using the tree as a roost for several days already, and would do so for a few weeks before they vanished. There was, it seemed, some kind of security in this small assembly, and in the particular landscape where chance had overtaken them. We walked on through the morning, making a wide loop through the hills, hoping that they were not travelling far from their roost, and that our paths would cross again during the day. They did. As we sat on a hillside for a midday lunch break the golden eagle of the group appeared not fifty yards away and lingered briefly just above our skyline. Could it be that for as long as this unlikely group of eagles roosted together, the golden eagle was influenced in its behaviour by the unwariness of the sea eagles? As the afternoon unfolded, questions crowded round like midges on a damp August evening.

Was it possible that because the new east coast project is based more or less due east of a sea eagle stronghold that evolved over years following the reintroduction off the west coast, it had inadvertently created the conditions by which young wandering birds from both coasts might find each other by exploring the same line of latitude and its great lochs and rivers? And given that the western half of that line – the Highland half – passes through one golden eagle territory after another, many if not all of them established for hundreds if not thousands of years, is it also possible that the passage of sea eagles through golden eagle heartlands might lure young golden eagles east along the same route? And if so, is there a wedge of land 130 miles long and a few miles wide that is being adopted by the young of two different species of eagles as a kind of thoroughfare between strongholds? If such a speculative theory were ever to hold any water at all, sea eagles would have to be making coast-to-coast journeys, but how could the Tay estuary population of young birds acquire awareness of the existence of a sea eagle stronghold on Mull? That may seem like the most baffling question of all, yet I think I know the answer: I think they find each other because it is in the nature of sea eagles.

Consider this. The young birds are taken from eyries in Norway with two or three chicks, and when they are between one and two months old. All they have learned from their parents is how to eat and what flight looks like. They travel by plane and van to the release sites in Scotland where they are held until they are strong enough to fly free and fend for themselves. Young birds from the original reintroduction programme that began on Rum in 1975 routinely wandered up to sixty miles away but a few made it to Northern Ireland and some to Shetland. Fewer still, responding to who knows what kind of inherited knowledge, made it all the way back to Norway. In 2011, a bird released in Ireland was found injured and rescued from a sea cliff five hundred miles away on the north-east coast of Scotland. I imagine that it too was heading for Norway.

The distances involved may be surprising, but the awareness of ancestry and how to track its spoor across ancestral terrain should not be. All kinds of wildlife tribes possess it. The most celebrated are humpback whales, which can travel from one ocean to another to find their own kind. I have seen more modest examples of it myself. When I was writing the first of my books about swans, *Waters of the Wild Swan* (Jonathan Cape, 1992), I talked to the staff of a swan hospital in Surrey where, at any one time, they may have up to two hundred swan "patients" recuperating on hospital ponds from illness or injury. They took one group of six fully recovered swans forty miles by road to release them on a reservoir. Their return journey was slow because of heavy traffic, and when they reached the hospital again they found the six swans had beaten them "home". They had flown forty miles across densely built-up terrain with no recognisable landmarks, and knew with absolute certainty how to reach their destination. And a badger sett I know in the hills of the Loch Lomond and the Trossachs National Park was "cold" for thirty years after the residents had been gassed. Then quite suddenly badgers returned several generations later. Nothing and no one took them there. But they followed the ancestral spoor. When I asked an older and much more experienced badger-watcher how he accounted for it, he shrugged and said: "Once a badger sett, always a badger sett," as if that explained everything.

And actually, it might. The fact that there was no sustained sea eagle presence in Scotland between 1918, when the last one was shot in Shetland, and 1975 when the first Norwegian chicks arrived on Rum, does not wipe from the bird's consciousness every sliver of awareness of thousands of years of occupation of this landscape, of cohabitation in this landscape with golden eagles, and (because it is much more flexible in its habitat requirements than golden eagles) the highways and flyways between Highland and Lowland Scotland. And what I saw one early morning above Loch Tay may have been nothing more than a parallel manifestation of the old badger-watcher's truth: "Once an eagle highway, always an eagle highway."

Its path is not difficult to track across the country, for the sea eagle may not need sea but it does need water. Flying west from the Tay estuary, it can travel all the way to Loch Tay simply by following the river. Or it can divert into the River Earn, which flows into the Tay near Perth, and follow that upstream all the way to Loch Earn, at which point it is only a few minutes' flying time and one watershed south of the pines above Loch Tay. Glen Dochart is the natural way westwards from both these lochs, and with the great land-mark mountains of Ben More, Ben Lui, and Ben Cruachan en route, the sea eagle can drift down Loch Etive to the Firth of Lorne and the Sound of Mull. It may sound fanciful, except that sea eagles have been seen in recent years – and continue to be seen – on every step of such a journey. They won't all be crossing the country, of course, but they are all meeting other eagles, and many of them will cross the path of golden eagles, and the numbers of such wandering sea eagles will only increase.

The implications are considerable, as considerable as they are uncertain, for this is new terrain, not just for sea eagles, not just for eagle-watchers, but also for nature itself. In 1975 there were roughly four hundred pairs of breeding golden eagles and there still are. In 1975 there were no sea eagles at all, and now there are a few hundred birds and an estab-lished and growing breeding population (fifty-seven pairs in 2011). It follows that it is a matter of time – perhaps a decade, perhaps two decades – before sea eagles outnumber golden eagles in Scotland, and much sooner than that they will out-number golden eagles in the islands and the West Highlands. But if, instead of human intervention in the shape of the sea eagle project, nature had contrived the unlikely circumstance that lured two breeding sea eagles to Scotland, and if the re-colonisation of Scotland by sea eagles was left solely to nature, the pace of population growth would have been a fraction of what has happened with human intervention, and I doubt if the golden eagle's domination would have been challenged in less than a hundred years.

So nature has a new situation to come to terms with, the introduction of a seriously significant predator, but in numbers and at a speed that it would never – could never – have countenanced. That situation is essentially man-made, as man-made as the extinction of the sea eagle a century ago. Nature will adapt, of course, because that is nature's way. How it will adapt, and how, between us, we will change the eagle landscape of Scotland, is the terrain of this book.

Chapter 3

THE WATERSHED

THE GOLDEN EAGLE GLEN ends abruptly in a wide headwall, lightly wooded with mostly birch, craggy, bouldery, and bisected by a white-knuckled burn whose procession of ragged waterfalls echoes far down the relative tranquillity of the glen. The best way from the watcher's rock to the watershed is to keep the burn's company. It was here one late June evening, with breeze enough to deter the worst of the midges and the glen softened by shadow after a long day of sunshine, that I toyed with the idea of spending the brief hours of pale darkness up on the watershed to watch the sunrise on the eyrie crag and see what unfolded.

Then the ring ouzel started singing. The song is full of jazzy rhythms and a tendency to belt out one haunting note again and again, like Sweets Edison used to do (if you know your jazz, Sweets is best known for his muted trumpet wiles filling in the spaces on the best albums of Sinatra, Ella, Tony Bennett). On and on, mellifluous and fluent, chorus after chorus, the song flowed like mountain burns, and then I had the notion that I would like to sit where I could see the singer, but without the singer seeing me. So I crawled away from the rock, my chin in the heather, one slow yard at a time.

The ouzel was in a low, scrubby birch tree by the burn and mercifully with his back to me. I crawled into the lee of a smaller rock, put my back to it and simply sat there. If he turned round I would be in full view, but I was dressed in something like the shades of the rock and the land, and I was silent and still and these things always help. Then the fox showed up. It was trotting along its accustomed path, one-fox-wide, but as it neared the ouzel's tree it slowed its pace then stopped. Then it sat. Then it put its head on one side. And if you were to ask me what I think it was doing, and if I thought you were not the kind of person easily given to ridicule, I would tell you that I think it was listening to the music, as I was myself. But now I was listening to the music while also watching the fox listening to the music while both of us were also watching the musician, who seemed to be oblivious to both of us. For three, perhaps four minutes, this situation prevailed, and a moment of my life was attended by the most enduring magic.

It all ended abruptly. The ouzel simply stopped singing of its own accord, and flew off into the deepening shadows of the burn. The fox scratched its nose with a forepaw, stood up, and wandered off. I sighed out loud, still under sorcery's spell. The truth is that I don't really know what the fox was doing, only that it seemed to be fascinated by the bird and the only fascinating thing the bird was doing was singing. Nothing in the fox's behaviour suggested it was stalking the bird. And as far as I could see, it was doing exactly what I was doing, nothing more, nothing less.

I know this though. If you spend a lot of time in one place with one overriding purpose centred on one particular species (in this case, the eagle glen and the eagles), you also acquire an onlooker's knowledge of at least some of the eagles' neighbours and fellow-travellers, just because you are out there and for long periods you are quite still and the neighbours and the fellow-travellers go about their worka-day business and you see at first hand how they get on with each other and how they treat you like a bit of the landscape. There are many days in this glen when you see no eagles at all, and a handful of days when they are rarely out of sight, but there are no days when you see nothing at all.

So each time I climb the burn at the right season of the year, en route to the watershed, I stop some distance short of the rock by the stunted birch and listen and watch and wait, just in case. And because I have gone often enough for long enough I now know that the ring ouzel is an unpredictable presence in the glen, that there are years – often several years at a time – when the bird is a regrettable absence, but sooner or later it returns.

Today, the day the pale eagle found me at the watcher's rock, there is no ring ouzel singing, and if the dog fox is out, I have not seen him. What I do know is that he no longer has any cubs this year, for the eagles took all three, one by one, and sometimes that too, is part of the story of how the creatures of the glen get on with their neighbours.

For as long as I climb the headwall, the only long view is behind me, back down the entire length of the upper glen, which has something of the feel of a steep-sided alpine meadow, and beyond which the descent steepens through forest and this burn becomes a turbulent river hell-bent on the loch, two miles distant and a thousand feet lower. Suddenly the slope relents, the confines of the headwall and the flanks of the glen vanish, and you breast the watershed.

You are drawn to the solitary cairn like a bee to a wild rose; the transformation in the nature of the land is instant and utter. Here, on the right kind of day, there unfurls such a breadth of heartland mountains cramming the horizon from east to west and folding away improbable distances north-wards, that you are aware of the catch in your breath and the uncertainty of your eyes as they struggle to take it all in. It doesn't matter how often I climb here, it is a place to which I have never grown accustomed. And sometimes, on the blue moon days, I have seen an eagle or a pair of eagles rise against the mountain horizon (an old voice in my head: "*You must learn to scan the middle distance!*") and on up into skies with-out limits, which is when and where you learn things about the eagles that you never see and never learn from a hide. They don't inhabit the world as constrained by the window of a hide or the lens of a camera. They inhabit the world, this world, this three-dimensional territory, with a mastery of airspace as utter as any creature that ever drew breath. Nothing is beyond them here: 100 mph in a shallow glide with wings half-folded and unbeating, a 3,000-foot free-fall that ends fifty feet above the ground and segues into a 1,000-foot power-climb then another freefall and the softest of soft landings on the frond of a rowan sapling. I have even seen the pale eagle rise on a thermal alongside a cliff and with her feet down and her wings wide, urging the wind to drift her purposefully – playfully – backwards through the air.

The watershed itself sends its infant burns south and north, and these, though they rise within yards of each other, are destined one for the Forth and one for the Tay and if their paths ever cross at all it must be somewhere around the Bell Rock lighthouse, fifteen miles out into the North Sea from

Arbroath on the Angus coast. The abrupt headwall of the eagle glen is behind and below me now in the south, but the watershed itself is a gentler land, wide and turned up at the edges towards east and west, and dipping gently to the north, where long, easy slopes slide away to one of the great east-west crossings of all Scotland.

So this watershed, this plain in the sky, is the tract of land that sustains the very heartbeat of the eagles' territory. I lived down there to the north at the foot of these long slopes for five years, and I learned a lot about myself and a bit about golden eagle territories during that time. It seemed to me that their territories had more to do with wind direction than physical boundaries, that there were, for example, negotiable overlapping areas between this territory and the nearest one to the east, because the birds themselves seem to prefer to hunt into the wind, for it affords them more flight control, so that if there was a west wind blowing the hunting birds from both territories would be out in the western reaches of their territories, and a tailwind would ease their burdened journeys back to the eyrie. At that time, both pairs prospered from a superabundance of rabbits on the lower slopes above the cottage where I lived, and on several occasions I saw birds from both eyries working the same hillside, which I assumed to be part of the negotiable terrain. But during the nesting season this watershed is the exclusive preserve of the pale female and her mate, and another eagle straying this close to the eyrie meets with ruthless intolerance. In late summer and autumn the boundaries relax. I have – once and only once – seen the four adults and the two single chicks they fledged lazily afloat on a warm late-September afternoon, the terrier-yaps of the young birds falling a thousand feet down the sky to my ears.

All six birds were high above these north-facing slopes where that brief rabbit infestation a decade ago now had perhaps established some kind of common ground and a precedent of mutual tolerance on the brief days of the wild year when the living is easy. The golden eagle's nesting season is a long one. By the end of December the courting overtures that reinforce territorial boundaries will have begun again, and so will old hostilities.

◎◎◎

I have climbed to the cairn after my long morning by the big rock on the floor of the upper glen and its five intoxicating minutes with the pale eagle; I have marvelled at the far-flung, wide-slung sprawl of mountains beyond; I have blessed my good luck and expressed my gratitude at being in such a place on such a day; and I have settled by the cairn, unpacked my lunch and my notebook, and unslung my binoculars from my neck and put them down on a flat stone where my right hand will find them. This is my idea of a working lunch.

I keep looking east across the watershed and south-east across the ridge where she disappeared. Occasionally I scan the whole sky through 360 degrees, then all the ridges down both sides of the glen, and then (and this is the hard part) the spaces between mountain walls. I don't know how many years it would have taken me before I tumbled to the fact that the place the eagle is most likely to be at any one time is somewhere in the space between mountain walls, but it was the nature writer (and golden eagle specialist) Mike Tomkies who told me: "You must learn to scan the middle distance." And of course it sounds obvious now but by the time we met he had logged many hundreds of hours of watching golden eagles from hides he would build himself from the raw stuff of the mountainside, and this piece of advice was one of the many fruits of his labours. The trouble with the middle distance is that it has no focal points, other than the birds that fly through it.

Mike was a friend for quite a few years before we had a pretty fundamental parting of the ways. We have not spoken since. But he had been generous to me with his time and his knowledge, and when he was sober he was good company, funny, full of stories from his past life as a showbiz writer, and an original thinker about nature. As I have already dropped one hint about my love of jazz into these pages, here's another. Whenever I think about Mike, I'm reminded of the tenor sax superstar Stan Getz (whose playing is simply my favourite noise) and his fellow tenor-player Zoot Sims's assessment of his multi-faceted and often troublesome personality: "Stan Getz? An interesting bunch of guys." Ten years after I last had anything to do with Mike, I still get asked about him.

☉☉☉

Late afternoon, and I think I might call it a day. It's an hour-and-a-half back down to the car from here, although, as I have ruefully pointed out to myself many, many times, about three minutes for a golden eagle that chooses to put its mind to the task. Vigils like this work up to a pitch, a plateau of intensity when my absorption in my surroundings is so complete that every sense is on fire and I am almost beyond the possibility of distractions. But it is not sustainable indefinitely, and eventually I start to notice that my mind is drifting, my eyes tiring. It's usually a good indicator that I have stopped being useful to myself, or in other words, time to go home. I stand, stow the signs of my vigil back in the pack, swing it on to my back, then swing the glasses through one last lingering patrol of the slopes, the sky, finally the middle distance, and there is that something, that sudden "?"

It is not observation so much as instinct.

What did I just see?

Where?

At first I cannot answer myself, but this is a phenomenon that has grown in me over half my life, this suddenly insistent voice of instinct, and I have learned to trust it utterly. There was something...

I retrace, as accurately as I can, the last long sweep of the glasses, as slowly as I can move them, up and down, side to side, all the way across those northern slopes, then when I see nothing I do it again.

Then again.

Then far, far down the mountainside I find the one thing that changes everything... another one of those rocks among hundreds of rocks that doesn't look quite right. There had been dark movement that stopped and settled as the glasses swept past the rock, then the frown, the question, then the re-examination of the landscape, then the rock, the rock that now hosts what I am certain (even at this distance) is a sea eagle. If it had been a golden eagle I would have stayed on the watershed, banking on the probability that its next move would have been back towards the heart of the territory. But a sea eagle is more likely to be wandering the big east-west valley with its rivers and lochs. So I hold a short discussion with myself about how I might close the distance to the rock without disturbing the eagle, and I opt for the course of a burn that slithers away through head-high heathery banks. The only problem now, of course, is that if I am to be unobserved as I travel, I will mostly not be able to see the rock, or the eagle if it moves. But the bird looks settled, and I have, after all, watched an eagle on a rock do nothing at all for four hours, so there is always that chance.

The going is messy, awkward, and several misjudgements of my left foot have given me a trouserleg wet to the knee. A cautious approach to a dip in the top of the bank should – I guess – let me look at the rock. However, now the rock is hidden by a small rise in the hillside, but perhaps if I crawl the fifty yards to the top, all will be clear? Halfway there, it occurs to me, a little belatedly, that if I could observe this sea eagle from the watershed, either of the resident golden eagles could have seen it from several times that distance and it is unlikely, to put it mildly, that they will leave it in peace.

I have still about ten yards to go to the crest of the rise when the huge raised primary feathers of an eagle appear against the sky dead ahead, followed at once by the rest of the bird – the sea eagle – travelling uphill at speed and no more than ten feet off the ground. My sense of what happens next is in the nature of slow-motion replay, which may be just my mind's way of accommodating the extraordinary sequence of events.

I freeze, on all fours, my left hand and left knee ahead of my right, and (I have no doubt at all) with my mouth wide open. It is this bizarre and quite possibly unfathomable creature that the sea eagle sees as it crosses the rise so close to me that its wingbeats sound like rhythmic gusts of a gale – *whoof, whoof, whoof* – that make the hill grasses tremble. The eagle banks at once to pass me on my left. For the second time today, I am staring into eagle eyes with the sun full on them. Then I remember the bunched fists of stowed talons, the simply colossal wings. Then clear sky.

Ah, but then there is the shadow. I see it topple over the edge of the rise and come straight at me, and considering how often I have symbolised the passage of the shadow of an eagle across a mountainside both in my mind and in print, and the fear and sense of subservience that passage must inflict on the mountain's lesser creatures, I eye this approaching shadow with a kind of dread. A collision is unavoidable. There is a split second – real or imagined – of profound chill, then the sun is on my back again and the thing is done.

Or not.

The crest is not done with me yet. It now unleashes a golden eagle, the pale female of the morning's encounter in pursuit of the sea eagle, but at the sight of me (and being a flier of an altogether more exalted class than its quarry), it soars on unbeating wings, gains fifty feet in a moment and drifts away east, still climbing.

All this, from beginning to end, has consumed less than ten seconds.

Still on my knees, I turn to look for the two eagles, the one last seen heading south-west, the other east. At first there is nothing at all, but then the sea eagle drifts by, having climbed a couple of hundred feet and turned away, heading north-west, where it crosses a hill shoulder and is gone. The golden eagle is circling, not a quarter of a mile away and slightly uphill from me. Then, just as she did this morning, she levels out and drives straight towards me, circles once, perhaps a hundred feet above my head, then drifts off west, and I wonder if in her own mind she has just etched a line of demarcation for the sea eagle's benefit — this far, and no further.

I stand, I wave to her, because it's a thing I do, and for no other reason. I ritualise my presence on her mountainsides, and she is accustomed to it.

On the crest, I find that I am closer to the rock than I had thought, so I wander down anyway. Then I realise that I actually know the rock of old, because it is not far from the house where I used to live and it was a particularly conspicuous feature of the landscape on what became my regular route up into the hills from the back door. In the five years I lived there, I grew accustomed to checking it out because the first time I scrambled up it I found a golden eagle pellet in a little mossy hollow on the top, looking like an egg in a nest, I had thought at the time. The pellet was an egg-shaped lattice of rabbit bones and fur. Those were in the days of rabbit plenty, where now there are none at all. Yet the wandering sea eagle chancing on this hillside homed in unerringly on what I was inclined to refer to as "Eagle Rock". Twice during one old autumn, I had seen the year's fledgling golden eagle from the eagle glen perch there in warm afternoon sunlight. But in the decade since I left that house the eagles have not prospered (for reasons I do not fully understand), most years they have failed to rear a chick, and I had not revisited the rock once. So coming on it from the "wrong" direction, I had failed to recognise it until the last moment. But now I see the clear ledge low down on its north side that was, from the first, an irresistible invitation to plant a foot and scramble the easy ten feet or so to its crown. I put my left boot there, push off, reach with my left hand for a small quartz outcrop, and the old routine sequence of hand and foot movements suddenly falls into place as if I had last swung up here yesterday. On the top there is a scatter of old bones… and a half-eaten piece of fish! The sea eagle, it seems, had carried it up from the river half a mile away from here, and that makes me wonder. There are any number of trees by the river and rocks on the slopes beyond its north bank that are much closer than this rock and the eagle might have perched on any one of them to eat, yet it flew half a burdened mile and crossed a main road to eat here. So perhaps the sea eagle has already spent some time here and learned something of the lie of the land. Or, it could be that as a wandering stranger here it carried its fish until it found somewhere secure to eat it, and recognised at once in the Eagle Rock those qualities that the local golden eagles have exploited forever, whatever those qualities may be.

Or… "Once a badger sett, always a badger sett," as my old badger-watching friend had suggested.

The idea fascinated me, and of course it made sense at once. And if it is true for badgers, then it is surely true for many more of nature's tribes. A particular landscape feature, or a habitat, does not stop being suitable for a particular species just because the species has either moved away or been wiped out. Likewise, the maps of Highland and Island Scotland are liberally punctuated with the words *Creag na h-Iolaire*, which in the Gaelic language that named much of that map means Eagle Crag or Eagle Rock, and which sig-nifies a landscape feature associated with eagles of one or both species. The fact that sea eagles have been gone from that map for anything up to a hundred years does not mean that they no longer recognise a *Creag na h-Iolaire* for what it is. They recognise it at once. And from Mingulay to these southern Highlands of my working territory, and now that I think about it, from the Cairngorms to Mull, it is a recurring theme of my working life that I constantly find common cause with nature on a conspicuous rock.

The growing Scottish population of sea eagles will likewise recognise not just a golden eagle when they see one but also a golden eagle landscape. Where that population is thriving, notably on Mull and Skye and elsewhere along Scotland's western seaboard, it does so in golden eagle landscapes. These young birds wandering through the heart of the country from either coast are demonstrably comfortable in golden eagle landscapes, have no fear of golden eagles, and it would appear that they even seek them out. The fact that this particular golden eagle persuaded this particular wanderer to take flight and leave its meal probably indicates a very young and inexperienced sea eagle and a mature and experienced golden eagle on its home territory. In Norway, where this sea eagle came from, and where the two species have lived side by side continuously (unlike Scotland with its century of sea eagle extinction in many parts of the country), it is the sea eagle that occasionally out-muscles a golden eagle from its prey. But I think this sea eagle, which had been compelled to abandon its fish on the very landmark I had christened Eagle Rock (and may not have been the first to call it that), was working with an ancient intelligence that identifies eagle landmarks within eagle landscapes. *Once an eagle rock, always an eagle rock.*

And now a new thought comes to me: in its flight from the river to the rock, just how close did the sea eagle come to the roadside cottage where I once lived? I had often seen golden eagles from the back garden, but never closer than about half a mile up the hill, never anywhere near the main road. The sea eagle works with different tolerances, and in Scotland's renewed acquaintance with the bird since rein-troduction began in 1975 – and especially since the east coast reintroduction began in 2007 – we are slowly coming to terms with the fact that the proximity of humankind is no deterrent at all. In the essentially Lowland nature of the east coast landscape into which the sea eagle has been most recently reintroduced, the natives are accustomed to buz-zards and occasionally ospreys, and anything bigger than that is usually a heron or a swan. A sea eagle on the roof or in the garden is shocking for all concerned. This, as we shall see, has led to some bizarre newspaper headlines.

My years in the cottage were far from the happiest time of my life, and only the wildness of my immediate surround-ings sustained me through it. I can only imagine with what ecstatic greetings I would have welcomed a sea eagle with a dripping trout in its talons dragging its shadow across my roof. Anyway, that was then. Now, as I turn my back on the rock for the plod back up to the watershed and down the entire length of the eagle glen, my heart is light, and my head is full of the coming years of new possibilities for my nature writer's work, and for the pale golden eagle that for so long has been at the heart of my portion of my native heath. Her world, and the world of many of her kin, is in the process of being turned upside down.

PART TWO

THE
WAY WEST

Chapter 4

THE TAY ESTUARY: THE SECOND BEGINNING

HERE WAS WHERE it began again, the second beginning. The first beginning had been far from here, on the island-splintered coast of western Norway between Bergen and Ålesund. If memory played any part in it at all, it would recall a bright ledge on an island cliff where parent eagles brought them food through the salt air, and the food like the air itself tasted of the sea. That place was loud with the shrill voices of sea birds and the deeper voices of the Norwegian Sea on the rocks. Their parent eagles flew in and out on wings that darkened the sun, and sheltered them when it rained. They knew even then that flight was the beginning and the end of everything. Their days and nights were rituals many thousands of years old, rituals designed to sustain the sea eagle tribe and to teach flight and hunting to newborns. But all that had ended in sudden, incomprehensible fear, with their parent eagles high overhead and screaming. That was the end of the first beginning.

⊙⊙⊙

There was darkness, and there were noises and smells that were not of the sea. There was a long, loud, shuddering, blind journey, a journey of flight, flight of a kind, but not of eagle flight. They knew it was flight, of course. They understood flight, why would they not? They were born understanding flight. They understood and felt the aircraft's lift, the angle of climb, then a banking turn from west to southwest, the levelling out at cruising altitude. They felt and they understood the aircraft's landing.

Then the movement and the noise stopped and the light returned and the sea was gone and they were among wingless creatures twice as tall as the tallest eagles, creatures that walked upright across the surface of the land, which is not their preferred way of travel. The journey ended here, and here was where it began again, the second beginning. It began again amid fear and the dark shapes of the wingless creatures whose voices were sinister whispers, and who left them food (but it tasted of the earth, not the sea). It was a quiet place folded away in low hills swooping steeply among small rough fields and woods. But here there was no sea-song in their ears and no nest and no cliff, and there were no parent eagles bringing food from the air, and the quiet and the shadows were as a cloak that this new land wears, a cloak of fear.

They were put in large cages, fenced all round and overhead, on an open shelf furred with whins and trees. The cages had perches, a shelter that was neither the overhang of a sea cliff nor the spreadeagle of great wings, and there was the earth-tasting food. There was no comforting presence. Their parent eagles had given them that, brought it to them on their sea cliff, unfolded it around them with the vast scope of their wings. Here there were no ancient rituals designed to teach flight and hunting to newborns. Instead, there were new rituals in which parent eagles played no part, so that when, at last, they fledged and flew free, they had to learn for themselves those skills of flight and hunting and wisdom their parents would have taught them from the sea cliff. So it took much, much longer for the newborns to learn how to be sea eagles.

But even then, even when they were released into this new country, they bore the imprint of the wingless creatures – plastic wing tags numbered and colour-coded so that they could be read through binoculars, transmitters stitched into their skin (a small aerial protruding from their backs) so that the people (being wingless) could track them. If you were to ask me what I thought about that, I would tell you that I think it is not nature's way, that if we are to reintroduce those tribes of nature our forebears once rendered extinct, it should be on nature's terms alone, that we have no right to put our fingerprints all over a young eagle. After all, the only ones of our own kind that we tag are criminals.

⊙⊙⊙

I grew up in Scotland when our only eagle was golden and Highland, the spirit of the high and lonely places as I saw it, a shunner of the people and all their works, our ultimate symbol of wildness since the people obliterated the wolf. But the sea eagle is demonstrably none of these things. Our much longer experience of Hebridean sea eagle reintroductions (almost forty years now) has taught us that this is a bird quite unafraid to come in among us, that it perches on rooftops, walks on beaches. It can be persuaded to perform almost on cue for boatloads of tourists with cameras (a thrown fish, a spectacular catch and flypast), and with a backdrop of ocean and island mountains the tourists take home with them unforgettable souvenirs. I have been on such a boat, on Mull, and at the end of it I was wondering whether nature was enriched or diminished by the process.

This land of the second beginning is no Hebridean fastness but the east coast fields and woods that fringe the Firth of Tay. The landscape is lowland and well populated by people. The city of Dundee, of all places, is just across the river. Why "of all places"? Because the city of Dundee is where I began myself. So when I stand at some midwinter dawn on a shore of the Tay estuary – any of its shores – and look around me, nothing and nowhere on earth is more familiar to me than this. This wider-than-a-mile river with my grey-brown home city heaped along its northern bank, those low, sheltering hills, that lazy midwinter sun reddening the North Sea horizon like a closed forge, the final daybreak flicker of lighthouses from Tayport to the Bell Rock… all this is the first of all my landscapes. It was here that I awoke into the world and fell more or less at once under nature's spell.

I know for example that any moment now the restless roost of thousands of pink-footed geese is about to erupt and climb in wave upon wave of flight and thrilling, meaningless noise. A few moments thereafter, a substantial proportion of the horde will align in the shape of a vast and uncertainly shimmering vee, and cross the low slopes of the blunt and wooded Balgay Hill, where the city used to end when I was a child. A compact and orderly flock of corporation prefabs used to stand in the shadow of that hill, ranged along paths at right angles to what was then the last street in town. One of those prefabs was the only building I ever felt truly at home in. It sheltered the first thirteen years of my life, at which point my parents were allocated a new council flat on the far side of the Balgay Hill and a kind of restlessness seeped into me then and has steered my life ever since.

The city of Dundee has expanded west and north and east in the sixty years since the sight and sound of geese over-head established itself as the earliest of all my memories. So I credit – and sometimes blame – the ebb and flow of geese (to and from their winter roost on the Tay at Invergowrie Bay and to and from the northlands of the world in spring and autumn) for who I am and what my life has become. And they still navigate to and from their roost by way of the treed flanks of the Balgay, no matter that there are a couple of extra square miles of rooftops now compared to the open fields of Hillside Farm that hemmed in the prefabs when they were built along one side of the last street in town. It is an ancient groove in the air that they follow between the Tay and the feeding grounds of rural Angus. For Dundee is a thousand years old and the geese were already there when the people came and sank their first roots in the good red earth, and for all but the last two or three hundred years they may well have had sea eagles to worry about as well as hunters. Throughout almost all my life, the nearest eagles to the Tay were golden and aloof, and they haunted – as they still do – the high hills of the Angus glens thirty miles to the north of Dundee. But since 2007, when the first Norwegian sea eagles were released on the east coast on Forestry Commission land in north Fife, all kinds of people and all kinds of creatures have begun to learn what all our ancestors knew, which is how to come to terms with such an unfurling of dark wings – and such a shadow! – in our midst.

And of course newspapers, radio and TV were primed to follow their every move, and especially their every misjudgement. They made much of one first-year bird, which on one of its earliest flights, blundered in on the unprotected free-range birds of a pedigree goose breeder. The owner, who happened to be a Church of Scotland minister (a fact that did nothing to undersell the story to the papers) disturbed the eagle as it was setting about a second goose, and there was an undignified scuffle between man and bird in the confined space of the goose enclosure, a thing that had not happened on the shores of the Tay estuary– if it ever happened at all – for more than two hundred years. In the process the man was cut, and of course he was persuaded to bare the wound for the benefit of eager photographers. MINISTER ATTACKED BY EAGLE! The headline was irresistible. It was a situation that a parent eagle might well have steered the hopelessly inexperienced and untutored juvenile away from, but the relevant parent eagles were two thousand miles away in Norway. Still, it is tempting to think of that small scar the goose breeder suffered as symbolic, as a kind of counter-strike by Mother Nature herself for all the wing tags and the transmitters and the aerials; the taggers tagged.

We will have to get used to this new definition of the word "eagle" in our hearts and minds as well as our neighbourhoods, fields and goose pens, and we will have to make our own accommodation with this extraordinary presence of nature, for nothing is surer than that the sea eagle is a bird that will ruffle feathers.

Just how extraordinary a presence? Consider again the impact of that aircraft flight on the eaglets' lives when they were still but a few weeks old, that alien and terrifying journey. Yet it seems they knew – precisely – their direction of travel and the course corrections along the way, *and remembered them*. As we have already seen, for as long as people have been moving between Scotland and the west coast of Norway to bring back sea eagle chicks in a sustained endeavour to reintroduce the species, some of those same birds, very occasionally and one at a time, have attempted to retrace in reverse the journey of the aircraft that brought them. One in particular made it back to Norway and stayed there to breed.

There is also this. By the time this new phase of reintroductions began, there was already an established west coast population. The birds spread up and down the west coast from Torridon to the Mull of Kintyre, but established strongholds on Skye and Mull in particular. Mull is due west of the Tay estuary. The first of the Fife birds to breed did not do so on the east coast, but on Mull. Like several other east coast birds it had found its way right across Scotland, meeting and quite possibly being reassured by other eagles along the way, following the lie of the land and the course of the Tay, and who knows what other signals and traits of inherited memory. It settled among its own kind, and in a west coast island landscape very similar to the one from which it was removed in Norway. Others have travelled to Mull and back and may well have lured Mull birds with them, and that too is in the nature of this eagle some of us now find on the doorstep. And what particularly commends those journeys to me is the astonishing coincidence that for as long as I have been responsible for determining my own travels around my own country I have sought out in particular and with ever increasing frequency the land between Dundee and Mull. The Tay becomes the sea just east of Dundee. It rises on Ben Lui, far to the west, and from the summit of which you can see at no great distance the hills of Mull. This, it seems to me now, is a destined connection.

⊚⊚⊚

Wherever I have travelled in the four decades since I moved away from Dundee and the shores of the Tay estuary, I have never quite succeeded in kicking the habit of leaving myself behind. I never stopped finding reasons to return:

Funerals, of course, including both my parents who are buried on that same west-facing flank of the Balgay Hill where the prefabs stood and where the dawn and dusk geese still throng the sky; and three of my grandparents are buried there too.

Work, of course, for there was a book, a restless prodigal son's homage to Dundee titled, almost inevitably, *The Road and the Miles* (Mainstream, 1996) and another about my grandfather called *The Goalie* (Whittles, 2004) – for that was his job, and Dundee's 1910 Scottish Cup win was his claim to fame. And there was a BBC Radio 4 broadcast from the top of the Balgay looking east to the sea, one of a series called *Letters from a Northern Landscape*, and after fifteen years there is still a weekly column in the city's daily newspaper, *The Courier*.

More recently, I came home again to research a book called *The Winter Whale* (Birlinn, 2008), the story of a humpback that turned up in the Tay in the winter of 1883–84, and its subsequent pursuit and protracted death at the hands of

Dundee's whalers, who happened to be laid up in port for the winter. Not long before I wrote the book, I had watched humpbacks on a life-changing month in Alaska, watched the four-inch eyeball of one whale as it nuzzled the boat six or seven feet below me where I leaned out over the rail, and I swear its eye purposefully sought out mine, albeit for a purpose I did not understand at the time. Perhaps, now that I think about it, nature was tapping me on the shoulder to show me my own world from five thousand miles away, for it was only after Alaska I discovered quite by accident that the Tay whale was a humpback. I had known its skeleton (for it hangs in a museum in Dundee) since childhood, when it haunted my dreams, but nothing more than the outline of its story and the McGonagall poem that immortalised it. But once I knew it was a humpback I found out that its story had never been written down, so I wrote it, and that one long-dead whale led me deeper into the world of whales, and the place of my own city and its river in their story.

Likewise, Alaska handed me a headful of bald eagles – I saw hundreds within the month, and in the same time, one golden eagle. The first bald eagle I saw was crossing a hotel car park in Juneau, the state capital. The second was stomping along a harbour wall about fifty yards away, the third through to the twentieth were standing beside their nests in tall spruces spaced out regularly as milestones as a Tlingit tribesman showed me the effects of clear-fell logging in sacred tribal lands from his boat, "ten miles of disrespect" he called it. The bald eagle is Alaska's sea eagle, the two species are biological kin.

So when I wander the shores of the Tay estuary today or clamber among its woods and low hills, I feel in the small of my back the urging of that unblinking whale eye the size of my hand, of that treetop eagle throwing back its snow-white head and carolling its territorial declaration for the benefit of the whole wild world. There are potent forces at work here.

Also, I miss the huge skies of the east coast when I am gone, the salt air, the sea views uncluttered by islands and the sunrises there. For if you are thirled by birthright to the east coast, your relationship with the sea is defined by the fact that the sun rises out of it so that it illuminates and galvanises the day ahead, whereas the west coast is a melancholic beauty of lingering sunsets and earthly sorrows, for yonder lies *Tir Nan Og*, the Land of the Ever Young, the afterworld, the Gael's definition of heaven. I miss Dundee's old stones and the speech and hospitable good humour of the natives, and of course I miss the majesty of its silvery river – McGonagall was right. I miss the bird hordes, the huge winter rafts of thousands of plump eider ducks, the myriad waders tribes, the terns and the Auchmithie puffins just up the Angus coast, the kittiwakes eternally chanting their eerie nom de plume a yard above the surf, the grey and the common seals lounging cheek by jowl on Tentsmuir's expansive sands; and in the last dozen years the bottlenose dolphins have returned of their own free will. But never, never for a moment of wildest imaginings did I conceive the notion that my place on the map would spawn a new tribe of eagles of all things. Yet here was where it began again, the second beginning.

☉☉☉

If looking for golden eagles in a Highland glen is like looking for a needle in a haystack, then looking for sea eagles around the Tay estuary is – relatively speaking – like shooting fish in a barrel. Just as in that curious way that the Highland landscape can make a golden eagle look small at anything other than close quarters (and close quarters is not a situation golden eagles enter into often or lightly), so the gently rumpled and essentially Lowland landscape of the estuary can make a drifting sea eagle look like a Zeppelin. Furthermore, and for the moment, the birds you are most likely to see here are very young, and young sea eagles being sociable creatures, you can occasionally see Zeppelins in unlikely numbers. The RSPB's director of the east coast reintroduction project, Claire Smith, has spoken in an interview of seeing *eight* juvenile sea eagles soaring above her near the release site. I can only imagine what that looks like, but once, far from here, I watched from a clifftop *one* young sea eagle take off from a beach (thereby removing from the beach what I had dismissed from my lofty perch as one more grey rock among many), swing out over the sea in a wide, climbing arc so that it drifted towards my clifftop into the wind, and about fifty feet above my head it held up against the wind and almost stopped, looking down. The bird was a dark grey silhouette against a dark grey sky on a dark grey and sodden day, but it still seemed as if it contrived to throw a shadow over me. The notional shadow of eight such birds in soaring close formation is perhaps the source of the primitive fear that impelled the tomb builders of Isbister, five thousand years ago.

But many of us will become accustomed to this kind of breathtaking encounter over the next decade and then far, far beyond, for such encounters are the sea eagle's way, and the sea eagle's presence in our midst will only grow.

☉☉☉

The grapevine works here too. The release site was a secret one, of course, although it was known to be on Forestry Commission land, which narrowed down the possibilities. I listened to straws on the wind, quickly ruled out the most obvious woodland, and started to look somewhere less obvious. This was, after all, the landscape of endless teenage bike rides, and many subsequent and irregular visits on foot and on two wheels and four, so I now harnessed a mountain bike and a good memory to my cause, stirred in a few prerequisite landscape features and my capacity for sitting still and inviting nature to come to me. There were false starts – there are always false starts – and I was turning everything from a buzzard to a big gull to a heron into sea eagles. So not quite like shooting fish in a barrel then? But you can get away with these things when you work alone and there is no one to scoff. Then I thought about that Loch Tay roost, and began to wonder where a handful of sea eagles might roost in the east, offering the possibilities of dawn and dusk spectacle.

I was driving up a quiet hill road late one afternoon. The landscape it traversed had looked promising from the first, but although it came with the grapevine's recommendation it had proved unproductive on a number of previous visits. This time, I saw a small space where I could run my car off the road onto a patch of grass beside a drystone wall. Beyond the wall there was a long narrow belt of trees, well-spaced and smallish oak, ash, rowan and holly. Beyond the trees a fence marked the edge of a field of rough pasture that dived steeply away downhill towards a snugly settled house with some small outbuildings. Beyond the far edge of the field was a wood of bigger trees, a mix of conifers and bulky big hardwoods, and beyond that a steep hill face with scattered Scots pines high on one flank, bare on the other. All this enclosed a small glen, as if a misplaced portion of the Highlands had wandered off when the land was settling down from the Ice Age, and having ended here it had shrunk to fit in with the rest of the landscape. In the changing light of the late afternoon it caught – and held – my eye for the first time, so that I wondered how I could have missed it before. I switched off the engine, felt coolness and an easy quietude flood in through the open window. I shouldered my pack, crossed the wall into the trees, paused by the field edge and let myself acclimatise to the place. That is easier than it sounds. It involves doing nothing at all other than being open to where you stand, and applying your senses to it. Then give it time. It's not Everest, just a little different from where I live and mostly work.

After a while I picked a careful uphill path just outside the field until I found a tree I could sit against in relative comfort, and which offered an unobscured view of the lie of the land. I thought it beautiful in a carefully controlled and contained way. Everything I could see bar the contours of the land itself had been put here by people to suit the purpose of living and working here, yet nothing at all looked out of place. If it was a landscape on an intimate human scale and fashioned into a human design, it had been adorned by nature. (I hesitate to say approved by nature but the effect had a harmonised quality that was easy to admire.) So what could be more appropriate than a carefully reintroduced top predator like a sea eagle to round it all off, as if it was the finishing touch of an intelligent design, even if the whole thing had unfolded haphazardly and over centuries, even if the way the landscape looks today is a consequence of many successive human occupiers putting their own stamp on the place piecemeal. Whenever people have manipulated a landscape, one of their first priorities is to assume the role of top predator and remove any natural rival. It occurs to no one to try and put nature's top predator back. Yet here and now, that is exactly what is happening. It is quite a moment in our own story, and in the story of this landscape.

Briefly, I daydreamed myself into that kind of work, living in that house down there, eagles for company, eagles that I had plucked from a Norwegian eyrie and escorted back here to this hand-picked place, watching the little glen turn through the circle of the seasons, narrowing my horizons to the intimate scope of the place. Then I decided I would rather live up here, in a new timber house dug into the hillside where the field fits the woodland corner like a hand inside a glove, and where the view would be my entire portion of the world. It was a pigs-might-fly moment, for none of that is in my nature, I'm forever switching landscapes, wishing I was on the coast when I'm land-locked, yearning for Skye from the Trossachs or the Cairngorms from Skye, that kind of thing. But the fantasy served its purpose, for it involved an intense scrutiny of the view, and that finally yielded a peculiar sense of movement on the bare half of the far hillside. What the…?

It was quite a long way off but I had noticed it without the binoculars, and now in the glasses I saw what it meant and smiled:

"Aha, rabbits."

The young sea eagle is a vagabond, a gypsy, a restless wanderer. But having established base camp, it tends to wander along beaten paths. When the Fife birds are in wandering mode, they mostly travel west along my notional eagle highway (and often back east again), or north and south up and down the east coast, doubtless puzzled by the sun that rises out of the sea instead of setting there. The ancient and — crucially — unbroken Norwegian sea eagle lineage into which they were born, has no east coast to speak of, and I believe that their instinct for many generations to come will be to seek out a sunset coast rather than a sunrise one. I think that mostly, they will go west and settle there.

But these birds have also acquired a home range here not a territory, for it takes around five years for a sea eagle to reach breeding age, when territory takes on a completely different meaning — but a reference point and a place on their idea of the map of the world, the place of this second beginning, a wooded hillside in north Fife not far from here, a place where they will quickly learn to exploit its every productive loch and hillside and goose-field, and the fish-and-bird riches of the Tay itself, a mile or two to the north. In that context, a bare, sun-facing hillside alive with rabbits is a perfect lure for the eagle-watcher. It was surely only a matter of time, and I had plenty of that...

...It was almost dusk and I was cold. I just knew that all over the north of Fife and the south of Angus and up and down the entire length of the estuary from Tentsmuir Point to Newburgh, people had been having — and marvelling at or cursing — unplanned encounters with sea eagles, while I, having carefully staked out my territory and satisfied myself that I had identified prime sea eagle hunting and roosting habitat, had chalked up a squad of wood pigeons, a skein of overhead pink-footed geese, and an inquisitive robin. A buzzard had loped across the rabbit hillside, which emptied of rabbits in an instant. Now, in the glasses, I could make out — just — that the distant hillside was aswarm with rabbits once more. The pines on their hilltop looked more inviting than ever, as if they had been added to the landscape just to accommodate roosting eagles the way we can make fake nests for ospreys and persuade them to nest there.

At this point, the balance began to tip in my favour. First, the patience of Job is a paltry force compared to my capacity to do nothing at all when I think it is in my interests. Second, I have long since learned to love the hours of fading light, of half light, of twilight, and the first creeping touch of the end of light. Time after time in my nature-watching years, that capacity has rewarded me with revealed secrets. It is the time of day when almost anything can happen, when the dayshift begins to shut down, and owls mobilise. Third, the moon came and stood above the hill and conferred magic on the hour and the land. Then they came in, directly above the Scots pines, and there were three of them at slightly different altitudes, working their wings slowly, wings like half-open parachutes, wings held briefly in stupendous down-curves, wings held straight and wide and tipped with primary feathers the size of paddles. With sea eagles, size is everything. There was a wedge of clear sky between the pine-fringed slope of the hill and the inky profile of a nearer plantation forest. I had been studying that wedge of sky for hours, pleading with its very emptiness, and in a moment there were three eagles flying there and the wedge of sky was suddenly devoid of space. The sudden impact of these birds on the sight is astounding, but in the oceanic, mountainous west the landscape feels as if it was born accommodating such creatures, as indeed it was, but in this little garden valley they seem to joust with the very hills for elbow room.

Chapter 5

THE TAY ESTUARY:
STRANGER ON
THE SHORE

THEY CAME ROUND THE HILL in an angled, uneven arrowhead, the middle one of the three slightly ahead of the other two, one above and one below. I wondered if they had any idea at all of the collective impact they made on the lesser creatures of the world, not to mention the watching mortal on the edge of the wood. There was next to no light left, the sky behind them was pale grey, darkening towards the hill edge, and they were as two-dimensional charcoal-drawn birds on that blank sheet of graduated greys. They were doing nothing at all, as befits birds wrought in charcoal.

I dared a fast sideways glance with the binoculars at the rabbits' hillside but it was folded away into cloaks of its own dusk.

Nothing at all was moving. Even the eagles seemed neither to advance or retreat, rise or fall, as if they were content with their new portion of the sky and might roost there on the wing, like swifts. A bit like swifts. The wind had fallen away from the trees and taken its voice with it. The greying earth inhaled and held its breath. It was a poet's hour.

Now fades the glimmering landscape on the sight,
And all the air a solemn stillness holds,
Save where the beetle wheels his droning flight,
And drowsy tinklings lull the distant folds.

Save that from yonder ivy-mantled tower
The moping owl doth to the moon complain
Of such, as wandering near her sacred bower,
Molest her ancient, solitary reign.

Thomas Gray's *Elegy Written in a Country Churchyard* is one of the landmarks of the literature of the British landscape, and simply my favourite poem, a work of rare beauty and layered depths, of precise, extraordinary, vivid language. My admiration of it goes beyond words, and over the years I have come to think of it as a poem wrought with a landscape painter's brushstrokes.

In the "solemn stillness" of the moment, then, three eagles had paused in the sky next to the evening hilltop I had thought might accommodate an eagle roost, then they drifted south across the face of the hill where they simply dematerialised against the fading light and the greying coalition of shadows. Where did they go? Did they settle in the valley? There is a small wood of hefty trees down there near where there must be a rabbit warren, handy for an eagle strike at first light. Or have they circled the hill and drifted up to its summit trees from below and behind? Have they settled there among the east-facing branches so that they benefit from the first rays of the earliest light? A little beyond the hill to the east is a small loch well stocked with fish, and the haunt of wild duck, geese and swans, and none of these are safe from the sheer weighted power of a hunting sea eagle, although the raw inexperience of some of the north Fife birds might make them think twice about swan or goose. But still, an exploratory dawn flight over the loch was just as likely as over the rabbit hillside.

I sat on for a darkening hour, wedged in against the woodland edge, looking as much with my ears as my eyes, although there is nothing like sitting dead still through the dusk to hone your night sight to something like the pitch of sensitivity we are all capable of. We have owl eyes if we work at night often enough for long enough. As it was, I lost the eagles in that low-lying depth of shadows, my eyes too poorly attuned for the task in hand. But once I wrote a book called *Badgers on the Highland Edge* (Jonathan Cape, 1994), most of the research for which was carried out in dark woods at night. I found my owl eyes then, and amazed myself. You have to keep using them of course, and most of us do not, myself included, but whenever I find myself in a situation like this, I wish there was more night life in my life. For a nature writer of all creatures, there are rarely more rewarding hours than the first and the last hours of the night.

The moon began to brighten the land despite its attendant cluster of clouds. It made the barn owl an easy spot, a shape-less patch of pale white adrift on the field edge far below me, but moving with that easy lope that characterises its hunting gait. I wondered if it might wheel about and come uphill, working the thicker grasses between the field edge and the trees where mouse and vole pickings are richer, but it clung to the valley bottom, silently working in and out of its shadows, in and out of moonlight, a heaven-sent moon for a hunting owl.

Curlews and oystercatchers began to drift inland, stabbing the air with their cries. A fox barked once... pause... again... pause... again, and was that third bark a different voice from within the wood at my back? If I was sitting on a straight line between two foxes, one of them might find me. Tawny owls laid round "ooh" vowels on the air, breathily soothing away the stab wounds, the barked harshness. The night was beginning to stir. A late tractor growled along the lane to the house behind two beams of light, crawled to a halt. The lights flicked out. The engine stuttered and cut. The night quiet deepened until it swam around me like oceans. And somewhere out there, up there, down there, or over there, three Norwegian-born sea eagles had settled into the deep shadows of a huge Victorian fir, or in the sparser shadows of the hilltop pines, or on a windless quarry ledge, or... and somewhere else, not far away, there must be others, and some of those were quite possibly among the pines that thicken the crown of the wood at my back. It was a strange, slightly eerie feeling, to consider the new possibilities of that scattered flock of Scandinavians, forcibly re-nationalised as Scots, each of them making its own accommodation with its new life in this new land in various and unpredictable ways. A handful of them (or perhaps all of them, and how would we know?) were responding to some handed-down awareness of how they might board that eagle highway that unfurled upriver and far into the west for unknown flying time, where they would find a place – a coast, an island, an ocean – facing the sunset that held essences of their original west coast home, the place of the first beginning. And there, they would find many other eagles of all ages, and golden eagles too (for they were familiar neighbours in their homeland). And these few, or these many will reconcile themselves to the West, and first to the highway that led there, and others who watched them leave would eventually be persuaded by their absence to follow.

These were the kind of thoughts I mulled over sitting in the wood above the field above the valley and below the hill with the scattered pines where I had thought a sea eagle might roost. This, I told myself, is a little different from the big rock in the eagle glen below the watershed, halfway along the highway from here to the island west. Walking back to the car in the moon-tinted darkness, I felt energised for the first time by the awareness of what had just begun to unfold here, of a wholly unpredictable journey for all nature, and I decided that I too would become the eagles' fellow traveller between here and the far end of the highway west.

⊙⊙⊙

A few days later, I drove back up the same narrow north Fife road, slipped the car into the same roadside stance, silenced the engine and doused the headlights, and let the last hour of the early morning half-dark rush into the open window. I sat still while the warm engine muttered to itself the way warm engines do when they start to cool. I wanted to nullify the intrusion of my arrival before I headed out for the same tree at the top of the field. I like to be at peace with nature when I go to work. It does not take long for nature to absorb a new arrival in its midst, however unpalatable it may sound and smell when it announces itself, and if it is instantly silent, still and unthreatening, that helps the process along.

I closed my eyes and listened. Almost at once I heard soft wings, small soft wings. I opened my eyes. A wren. It was on a branch of a skinny little holly bush, about two yards away and at eye level. A wren at eye level is disconcerting. The eye is perfectly round and perfectly black and perfectly tiny. And there is only one eye because it looks at you side-headed, so you find yourself wondering if the eye you can't see is watching something else. Then I remembered the whale eye, four inches across, and how I had taken it for nature's messenger. And here was a wren eye – what, a millimetre across? – at more or less the same distance. "What message do you have for me?" I asked it softly.

The irony is potent. I am looking for eagles. I find a wren. In this landscape, the wren is the eagle's fellow traveller, and that inscrutable little full stop of an eye is as much nature as the eagle's yellow glare. And then, it occurred to me that this was not the first time in my life that eagle and wren had aligned in portentous circumstances. The following is from a book I wrote in 1990, *A High and Lonely Place* (Jonathan Cape). The incident happened in Gleann Einich in the Cairngorms, and in winter.

Black clouds massed over the Moine Mhor, that high plateau that sprawls away from the top of Gleann Einich's cul-de-sac headwall. The sun fitfully blazed and snuffed out daring spotlights so that the glen's mountain walls were never still, never uniformly white, but every subtle shade of dark white, pale white, and sullen grey.

The river had begun to narrow yard by yard as the strangle-hold of ice and heaped snow banks encroached. I paused to watch the light's dance, decided on a brew, and as the stove wheezed I became preoccupied with a dipper's unflinching zeal, perching on an iced rock to sing, swimming and diving down to feed after its own perversely amphibious fashion on the riverbed. Is there a grittier gladiator-of-the-wilds than the dipper, I wondered, at which point there was a tiny scuffling inches from my feet, and a determined busy-ness under the overhanging lip of the snow's newly redefined riverbank. It was a wren.

There are times when nature's logic is incomprehensible. There is no common species of bird in all Britain more suscepti-ble to winter harshness than the wren, no winter climate harsher in all Britain than the Cairngorms. We – the wren and I – were three miles out beyond the pinewoods, both of us heading upstream, the wren foraging with some success in the very jaws of winter. Wrens fend off the worst of winter in communal roosts, and there are many records of wrens packed into tiny improvised shelters for warmth, including the tragic failures – forty found dead in a single nest box. It seems nothing more than a tidy way to die. So why this single bird speck on the winter face of the Cairngorms?

The coffee brewed, the wren busied on upstream, the plateau wind whipped away the towers of cloud and the sun won an unfettered hour. I was still turning over the wren conundrum, wondering what it feels like to delve into those icily sodden overhangs and pluck a spider from the white darkness, when my dawdling glance fastened on the unquestionable silhouette of a golden eagle high above Carn Elrig. It flew south-west out over the glen on unbeating wings (the wren's were a restless blur), then launched into the matchless routine, the roller-coaster dive-and-climb of the male eagle's display flight that would tilt the whole Cairngorms landmass into yawing skylines, the wing-folded free-fall then the turbo-charged climb power-driven beyond the comprehension of mere mortals or the consciousness of wrens. The performance marked that critical point in the wild cycle of eagle life that announces an end to the old year, the bird convinced that despite the day's snows and the river ice that winter has frayed beyond repair. A commitment is launched to a sustained expenditure of energy that will last half a year, beginning with this stylised sky-dance.

Eagle eyries can be immense. The rigours of construction, then mating, brooding, rearing the eaglets, killing, fetching and carrying pray as large as a dog fox or a deer calf or as meagre as mouse or vole on the lean days, teaching the young the phenomenal skills of eagle flight… all of it in the face of the harshest mountain climate in the land, all of it a routine for which only eagles are fitted. After the encounter with the wren, this new beginning served to demonstrate something of the scale of nature's repertoire which is required to fulfil the demands of the ecology of a single mountainside, a repertoire that extends from the eagle to the prey of a wren.

That same late February day, the wren was hunting down scraps of survival three thousand feet below the celebrating eagle. And even then, a hard week of frosts or prolonged snow could still kill off a tiny bird so weakened by winter. Yet when winter relents sufficiently, the wren may well nest higher than the eagle. There is nothing rare about wren nests at two thousand feet, even though they may be more familiar in your back garden. The eagle, especially if it is a tree nester, often nests several hundred feet lower here. So there will be time when a boulder-singing wren pauses in mid-chorus to look down on the formidable

*back and wingspan of a golden eagle bearing home a haunch of
deer carrion or a brace of ptarmigan…*

North Fife is not the Cairngorms, neither in its landforms nor
temperament, its new eagles won't cruise a mile high, at least
not here they won't, and there are no mountains to challenge
the tenacity of pioneering wrens. Otherwise, however, the
same principles are in place, the politics and the practicali-
ties of life at the top and bottom of the food chain are much
the same, and the ecology that makes a valley like this tick
is as complex and infinite in its reach as Gleann Einich. The
wren and I watched each other for perhaps a minute, which
is a long stillness for a wakeful wren (though perhaps not for
one that has just been rudely aroused from a pre-dawn doze).
What disturbed our trance-like moment was the blackbird's
arrival at the other end of the wren's branch. The wren dived
down into deeper cover and was gone, the blackbird pro-
tested at something beyond my reach and vanished in the
opposite direction. The holly branch was quivering and bare.

Time to go. Fifteen minutes later I was back at the top of
the field where I settled once more, this time to watch the
valley empty of darkness and fill with light, and to resume my
pursuit of its eagles. "If," I told myself aloud, "they are still
here, that is."

I was never so enthusiastic about dawn vigils as dusk ones,
for no reason that I can put my finger on, and that despite the
fact that some of my pre-dawn rises have produced moments
of wildness that are still vivid twenty or thirty years after the
event – an osprey standing on the edge of its eyrie near the
Lake of Menteith to scatter rainwater from its plumage into
the glare of a dazzling red sunrise, a barn owl that almost

flew into my head at around 3am one midsummer morning, a merganser that swam past my feet without knowing I was there a yard away on the bank of its burn, an otter asleep, a skylined red deer stag that posed beside a setting moon, a wildcat on the prowl... these and other encounters punctuate the near and far contours of memory. Dusks, on the other hand, have more to do with landscape moods than memorable encounters, but in all my dusks, and for that matter all my dawns, there have been many eagles. So I sat again, poured coffee, cupped hands round it, and waited, watching the sky's paint dry.

The "paint" began to seep in from the same wedge of sky between forest and pine-treed hilltop where the three eagles had materialised those few evenings ago. It was more or less due east of where I sat, and it began to fill hopefully with pale yellow and from the bottom up, so that it widened as it rose, a stain that spread south across the hilltop and between the pine trunks and tiny spaces between branches, then beyond the trees and down the far side of the rabbits' side of the hill. There it lingered long enough for me to register the presence of a small horde of rabbits all across the lower slope. Surely the eagles would home in there as soon as the light brightened.

It dulled. The liquid watercolour yellow was overwhelmed by a shuddering wash of light grey travelling at twice its speed, a shroud of high cloud that precluded the spectacle of a sunrise so that dawn stole furtively round the edges of the hill instead, nibbling disconsolation around the edges of my expectant mood. So I raised my glasses to the high pines, aware that I was suddenly trying to convince that part of me that suddenly needed convincing that they offered the best prospect of early sightings, because (I reasoned with suddenly less than flawless logic) if I were a sea eagle, the high pines were where I would have roosted. At this stage of my slowly accumulating experience of watching sea eagles, I had seen three roosts – the one above Loch Tay and two on Mull, and they were all on trees with an open outlook. Here were trees, not just with an open outlook but also rabbits for breakfast, not to mention that glimpse of three eagles one evening earlier in the week. And here too, a frisson of doubt perched on my shoulder and started muttering in my ear. I flicked it away with one of my stash of convictions that I keep in a pocket in the back of my brain for just such a purpose, determinedly refocussed the glasses on the trees and the hilltop, and drifted the glasses optimistically out into the wedge of sky from time to time. Meanwhile night fled, dawn became daylight and the workaday world kicked in.

⊚⊚⊚

I felt rather than saw or heard the sea eagle, for it seemed to manifest itself moments before I saw it as a disturbance of air, low and to my right and (as I remember it now) too far behind my right shoulder for any kind of physical recognition to have been possible. There was a bellowing crow almost in my ear, then a second crow not far above my head, and it was clear to me before I saw them that they were furiously on the move. I turned at the sound and whatever awareness that had preceded it, twisting awkwardly round my right shoulder (that routinely stiffens at the slightest excuse), to see a sea eagle twenty yards downhill, ten yards inside the field, two yards off the ground, and swatting the air with gulping wingbeats the size of fireside rugs. The tail was startlingly white in so many shadows. The crows homed in on that tail from either flank as if it was a target being towed along by the eagle for the purpose, the nearer one close enough for me to hear the creak of its wings, the further one diving down, a yard behind the eagle and closing.

Then there was a moment of connection. The eagle, besieged by his black tormentors, suddenly looked upwards and left, and made eye contact. God, the whale again! My head and shoulder movement to confront the crow noise were more than enough to betray my presence. As with the whale, I now have an image of that sea eagle pinned to the inside wall of my skull for my brain to look at whenever it needs one. The eye – the one side-headed eye like the wren – is pale yellow, its shade prefigured in that first dawn flush. It is darkly hooded. It sits high in the profile of the head, which is lightly mottled pale grey and fawn, and just astern of that massive yellow hooked slab of a beak. You reach for the word "massive" a lot with sea eagles.

Memory has fixed that eye just ahead of the "elbow" of the left wing at the moment the wing reaches the lowest, widest reach of the downstroke. The "flying barn door" cliché that so many media outlets insist on dropping into every mention of sea eagles is but one perception of the bird, as seen by craning necks and heads from far below. (In truth, it was a phrase carelessly used by a crofter in conversation with naturalist Roy Dennis, which Roy subsequently quoted in an interview, and which has since become the inevitable media shorthand.) But the bottom of the downstroke at close quarters and from slightly above reveals a bird cloaked in a corrugated drapery of up-curved feathers, folds and folds

of feathers of such a size that they redefine the very word
"feather" in your mind. No part of what I can see looks like
a bird shape, but the whole thing rather presents the form of
a bluntly cornered and tilted triangle, the wing elbow at the
apex, the tail feathers at the right corner and the primaries of
one wingtip at the other. Only that eye, forehead and beak
protrude beyond the smother of feathers.

The upswing begins, the head vanishes, the triangle col-
lapses and the moment is done. In its place there is an articulate eagle, low and unhurried over the field with two crows
in thankless pursuit. The eagle contrives a mid-air convulsion
that flips the whole, improbably nimble mass on its back so
that, instantly, the crows are confronted not with a white
tail to torment but with a raised pair of eagle talons. It is
enough. They wheel away in formation and climb back to
the trees at the edge of the field where they try to restore a
semblance of dignity. The eagle reverses the convulsion and
resumes its low-level flight, which appears to have rabbit as
its destination.

Such is the joyous unpredictability of working with young
eagles in a new landscape. We – eagle and eagle-watcher –
are making it up as we go along, and then revising what we
have made up almost as soon as it is made. I had made my
best guesses based on my frail grasp of sea eagle logic, not
quite knowing whether such a thing even exists. I had settled
on a particular corner of a particular landscape based on the
little I thought I knew, only to have a sea eagle ambush me
from behind. I watched it fly towards the rabbit field, but
about two hundred yards short of the field it banked and
climbed and perched high in the open edge of that small
plantation of big conifers. Then it began to preen, with a

settled-looking I-may-be-some-time air. I scouted with the
glasses all over the sky and in every corner of the land that
was available to me from where I sat, but the preener in the
tree was the only eagle in sight. So I decided while I waited
for events to unfurl that I would try and reconstruct the
sequence of events that had just unfurled behind my back.
The trees between the road and the field, and which were
immediately behind where I first saw the eagle, were too
awkwardly spaced and too thickly branched for such a huge
bird to have flown through them, yet it was surely too close
to the trees and too low to have flown over them.

Two possibilities occurred to me. One was that the eagle
had flown up the edge of the field and I had not seen it simply
because I was facing the other way. But the crows had only
given voice at the very last minute when the eagle was already
very close, and I know from many years of watching golden
eagles that crows are smart and sharp-eyed, noticing creatures
that will travel hundreds of yards to harass them, hurling abuse
the whole way. It was clear that here the crows had travelled
no distance at all to greet the sea eagle, and that could only
mean that it had astonished them as it astonished me.

So the second possibility is this: the eagle was there all the
time.

It was there, perched in the trees between the road and
the field, and facing across the field to the hill. Perched in the
trees when its head swung round to glare at the nature of the
intrusion represented by my car engine and headlights, relax-
ing again in the silence that followed when the sound and the
lights died. But then there was the blackbird alarm; alarmed
at whatever it was I failed to see – eagle?

Perched in the trees when the car door closed, not loud,

but deafeningly conspicuous at that moment in that landscape, the sound placed by the eagle precisely where the car engine had stopped. Perched in the trees and glaring at the place where I stepped beyond the fence into the field edge and began walking uphill towards its very tree, adjusting its head position minutely and moment by moment as I came closer. Perched in the trees when I walked *beneath* the very tree not looking up but looking out across the field to the silhouetted hill with its pines believing *those* distant trees might harbour eagles. Still perched in the trees when I reached the high corner of the field and turned right, hugging the fence, taking pains to stay with the shadows so that I would not reveal myself to eagle eyes that might be scanning the dawn field.

By now the eagle's head had swung through 180 degrees, and it was still perched in the trees when I sat and grew still, and it decided I posed no threat, and that at least until daylight came, stillness would serve its cause best. When daylight did finally arrive the crows spotted the eagle and urged it on its way, but not perhaps until it flew, and it had fooled them all that time with its tree-coloured stillness. But its sideways glance up at me as it flew confirmed its first appraisal: no threat.

If it seemed to me to be extraordinary behaviour for an eagle, it was only because the eagle with which I am familiar is not this one, because a golden eagle would not be seen dead in Fife – even north Fife. And because in those few and far-flung corners of Scotland where golden eagles can be seen from roads they almost never perch close to them, and the odd exception that proves that rule would be a distant blur heading for the horizon by the time any vehicle driver with local eagle knowledge parked, opened and closed a door, shouldered a pack and turned to look around.

⊙⊙⊙

There is a sense of thrill in the back of my throat as I stitch together the threads of that second possibility. As much as anything else I can think of, it is like the visitation of the whale, the eye contact housed in such outsize spectacle. This stranger on the shore of my boyhood and youth has not only turned my head and forced me to consider the place of eagles – all eagles – on the historic map of my country; it has also beckoned me home and out of the Highland surroundings where my life had dropped anchor so that I might consider the new prospect of eagles in the place where my childhood self once looked up at a skein of geese and took nature's hand for the first time. The wonder of it was scarcely containable.

And yet, and yet, these same sea eagles that had begun to enthral me were also making enemies with axes to grind. The old morbid loathing of anything with a hooked beak is surely our most stubbornly enduring Victorian legacy, and its disciples have never had a hooked beak quite like this one with which to vent their spleen and grind their axes. The Scottish media, which rarely reports wildlife stories without either throwing up its hands in horror or making trite jokes, seems to delight in their anger uncritically. The sea eagle that attacked the minister just over the hill there is a case in point. It later emerged that he was not attacked by a sea eagle, his prize-winning rare-breed (and unprotected!) geese were attacked and he had cornered the bird in the wooden goose-house and inadvertently blocked its exit, so as it tried to escape he was an obstacle the sea eagle tried to do something about. He conceded that he eventually subdued it by standing on its wings, so he was hardly a victim here. Then there was a farmer near Perth who showed off a dead lamb to a newspaper photographer, telling the world that it had been killed by a sea eagle. Tests subsequently showed that it had not been killed by a sea eagle, but by then the original headlines had done their work. And then there was the woman who witnessed a mid-air attack by a sea eagle on a swan, and without stopping to consider the possibility that such a spectacle is as old as swans and sea eagles, chose to rage to a reporter that "these birds are killing on a whole different level now", adding what has become an increasingly frequent refrain, "what's next – a child?"

And so on and so on, and the anger grows and rumbles in the fields and low hills around the Tay estuary like thunder. Hysteria is so much louder than wonder. It serves to underline more tellingly than anything that has happened here since we killed off the last wolf, how wide the gulf has become between our everyday lives and the everyday lives of nature, how impenetrable our lack of understanding has become since we annihilated the big predators. And now that a new predator has been restored to a landscape where we never expected to see it because most of us never knew it had ever existed here, the truth is that we don't know what to do with it or how to respond to it in a way that is relevant to the twenty-first century rather than the nineteenth. It disappoints and depresses me.

Then suddenly an email from a friend arrived about the day she met her first sea eagle on Tentsmuir beach, which is north Fife's most north-easterly corner, a spectacular place of pinewood and sand dunes, seal colonies and sea birds, and a seaward view as wide as the sky and a sky the size of the North Sea. Ann Lolley works for a small community-environment trust and I had just done a little bit of work with her in Dundee. I liked her attitude, and when I heard about her sea eagle I asked her to write it down for me, thinking she might have something interesting to say about it. I got rather more than I bargained for. This is what she wrote:

On a rather grey summer afternoon, characteristic of much of 2012, I decided to leave work early and go for a walk along the beach at Tentsmuir. I don't know what nudged me into taking this decision since this was an extremely unusual thing for me to do. However, Tentsmuir is a favourite place for me and when going there alone I choose times when there are unlikely to be many people. That grey weekday afternoon there were only one or two cars scattered among the trees.

I parked between two Scots pines and, as usual, set off over the dunes and straight to the beach. I must admit that sea eagles were not on my mind at that very moment. However, I had been speaking to a friend about them earlier in the day, saying I still hadn't seen one of the Fife birds, and was he sure I would know one when I saw one? He assured me I would.

I stepped from the dunes to see the familiar wide expanse of beach and sea, and there, right in front of me and barely five or six metres from where I stood, was a large bird standing stock-still. It was looking at me in profile with one very large clear eye; a huge hooked beak, closed wing, and touch of white tail were all visible.

Instinctively, I stood still, expecting him to fly off the moment he saw me. He didn't. We stood still, simply looking at each other, somehow timeless, present; my mind had been stopped in its tracks.

Who knows how long we stood like this? It struck me then that the bird had no fear of me, and that I, rather than seeing him, felt his presence from a place deep within me, somehow absorbing and knowing his essence. I heard my inner voice say, "Hello, you must be mister sea eagle," and it was like the bird was also acknowledging me. The sea eagle and I were greeting each other with "Namaste, the divine part of me welcomes the divine part of you... and likewise the other." [1]

It was a meeting similar to that "knowing" when feeling a strong connection towards a person you have only just met, no need for words; somehow you know that they know something of you, that you hold a part of them in you. My sense is that such situations are always mutual even though we do not often pluck up the courage to acknowledge it to ourselves, let alone to the other, be it bird or man.

The size and perfection of the bird and its one clear eye which held contact with mine seemed to mesmerise me and hold me in a state of wonder. Somehow I understood deep within myself that there must be a reason for our meeting. This was an offering. He was saying: "If you just slow down and take more time, then these are the things that will be revealed to you; they are things worth more than a thousand precious jewels."

Then just as easily, he opened his wings and took off, the white tail now very visible as he headed off to the west. He was in no hurry as he lifted into the sky, everything happening in slow motion.

As he left it was like something shifted for me and I was

---

[1] *Namaste:* In the Indian subcontinent, a gesture of placing the hands together at the heart *chakra* (point through which energy flows), closing the eyes and bowing. The word literally means "I bow to you". In the West, the word is usually spoken; in India, the gesture suffices. It acknowledges the soul in one by the soul of another.

released from my stillness. I moved quickly to the spot where he had stood, expecting to see clear, huge, sea eagle footprints or at least claw prints in the sand. Had he left something tangible behind for me? Nothing remained. Just the image in my mind's eye of his eye.

As I continued my walk along the beach, my mind now back in action, I realised that I'd thought my first sea eagle experience would be of "the barn door in the sky", not an eye-to-eye contact on a beach where I felt that if I had moved quickly I could have caught him in my arms.

Thinking back, 2012 was for me the year of two other intimate encounters with creatures of the natural world. Around springtime I had met a hare in my garden, no more than a couple of metres away. We stood watching each other over my rhubarb patch for several minutes, him cleaning his long ears and me thinking, "Is this really a hare, and why is he hanging around here?" Later in the year while solo kayaking round Holy Island off Arran, I spotted a rather large fin and tail: I was being accompanied by a basking shark on my paddle around the island.

It was a year for me of closely connecting with some other inhabitants of the earth, air and water, seemingly through their choice, not mine. Perhaps such species have the role of reaching out and mesmerising us enough to make us change all of our destructive habits that impact on the earth. If we truly begin to know that we are connected to everything on this planet and in the universe, perhaps we will change our actions. Maybe we will become the final element that will bring about change?

Such testimony does not make good headlines. Its passive, internalised nature cannot compete with the uncensored ravings of the "threat to our children" school of thought in our headline-hungry, Internet-obsessed climate. So you might think that any hopes that the stranger on the shore might be accorded a hospitable welcome in the wider community and a thoughtful one in the media is more pie-in-the-sky than flying barn door.

And yet a woman who was not even sure she would know a sea eagle if she saw one was moved by what was a pretty startling encounter not to fear or rage but to stillness, admiration, and a sense of connection. The trouble, from the sea eagle's point of view at least, is that people who do make that kind of connection with nature usually experience it quietly and alone and lack either the opportunity or the inclination to articulate it. If they could, I believe that many people would marvel at it, and nature would win many more friends. None of us knows, incidentally, how we might respond to the circumstances Ann Lolley encountered unless we encounter them for ourselves. I certainly don't. I have never had a standing sea eagle stare me down from a few yards away then entered into a partnership of stillness with it. Having read Ann's account, I can imagine how I might respond, but unless and until it happens, I won't know for sure.

More importantly, the point is that for every irate headline that sea eagles inspire around the Tay estuary as they try to find their feet in what is still a strange land, there are dozens of unrecorded encounters and sightings, for the people who have them do so in states of unreported indifference or fascination or rapture or something in between. It may yet be that in the folk mind the sea eagles out there on the cliffs, the evening tree roosts, dicing with the wind turbines in haar and low cloud, drifting down to torment the eider duck rafts on the Tay, lifting a trout from under the noses of the anglers on Lindores Loch or Loch Leven, or standing tall and motionless like a grey-brown lighthouse with a single sunlit eye on the sand at Tentsmuir... it may be that the folk mind is quietly coming to terms with them and learning if not to love them at least to admire them and to find a new respect for nature's audacity in what to most of us is still the most unlikely of landscapes for eagles. Far in the west, at the Atlantic end of the eagle highway, the protests are the exception to the rule now. Mull announces itself to the world as the Eagle Island, and nature has won all the arguments.

◎◎◎

At the top of the field facing the hill, I stirred stiffly and stood to ease limbs from their long stillness. The sun had never quite got the hang of the early morning and retreated into high cloud. The eagle that had treated me to a fly-past had been more or less motionless in its tree for around two hours now. Then suddenly it was airborne. I had stood up without the binoculars either in my hand or round my neck, and in my haste to find them I stood on them, smearing mud on the lenses. (I have never been a great respecter of technical equipment, and sometimes the technical equipment bites me back in revenge. Binoculars fit comfortably within my definition of "technical equipment".) I wiped them on the inside of my sweatshirt, which was the handiest surface for the job, while trying to keep the eagle in sight. I lost it briefly against the low ground then found it again, much higher and against the sky. Just as I was wishing I had seen what must have been a spectacular power climb (I was guessing – I have

seen golden eagles do it many times), I caught a new movement low over the rabbits' hillside and realised that I was
now watching two eagles, one with a white tail (now clearly
visible in the more or less cleaned glasses) and one, much
higher, without a white tail. Where had that come from?

I watched the rabbit hunter, saw its flight steepen then
level out then slow abruptly as it adjusted to the rabbit slung
beneath it. The fluency was impressive, the rabbit was simply
plucked from the hillside in much the same way that this
eagle tribe takes a fish or a duck from water without getting
its feet wet. Then I watched the higher bird cruising in wide
circles. I saw it descend slowly, awkwardly, and with none of
the other bird's fluency. Two rushed passes across the hillside
produced nothing at all and cleared the place of rabbits, and
the eagle perched in the very tree where the white-tailed
bird had sat for two hours. It struck me then that, deprived
of the opportunity to learn from a parent bird, this younger
eagle had been watching the adult with the white tail, which
is the sea eagle's badge of maturity; may even have been following it around to watch and learn how it did things. One
way or another, nature finds ways to feed itself or perishes in
the attempt.

I called a halt to the early morning shift. I walked back
down the field edge to the car. I was stowing my pack and
boots when I saw two sea eagles about fifty yards apart and
heading south. I watched them through the binoculars; one
with a white tail, one without. I lost them after about a mile,
but because I know what lies in the south that might interest them, I decided to follow by car, which is not quite as
preposterous as it sounds. Half an hour and several distant
glimpses of the pair later, I was in a hide at the RSPB's Vane

Farm reserve at Loch Leven watching two sea eagles drift round a shoulder of Bishop's Hill, one with a white tail and one without. The thousands of birds that throng the reserve and the loch have always had wildfowlers with guns to worry about, and foxes and peregrines, but now the stakes have been raised, and now they have those new and fearful shadows darkening their sky.

I am still no wiser about where they roosted, but I'm not done looking yet.

Chapter 6

THE HIGHLAND EDGE

THE FIRTH OF TAY is at its widest just west of Dundee – two miles wide between the villages of Kingoodie on the north bank and Balmerino on the south. It tapers inland to less than a mile wide at Newburgh ten miles upstream, and to no more than a few hundred yards wide when its course bends north-west then north through the city of Perth. If your destination is due west from the sea eagle rearing-pens of north Fife, following the course of the river beyond Perth puts a mighty diversion on your journey, all the way north-west to Ballinluig then west to again to Loch Tay, Killin and Glen Dochart. On the other hand, if you are a first-year or second-year bird you don't know where you are going anyway so what difference does it make? There is evidence aplenty to suggest that some young sea eagles do just that, or cut the corner to Loch Tay by crossing the hills at Glen Quaich to Kenmore, where – who knows? – they might fall in with a young golden eagle roosting in a high pinewood.

But there is a more direct alternative – the lesser water-course of the River Earn which meets the Tay just west of Newburgh, and heads more or less west along its entire length. If, as I suspect, it is the quest for a Norway-esque coast of mountains and islands that pulls many a young Fife-reared eagle west along some kind of coast-to-coast high-way, then they surely find confirmation here in the way the course of the Earn heads upstream to country that rough-ens and rumples and wilders as it westers, until it submerges itself deep into distant mountains. From time to time, and throughout the story of the east coast reintroduction project, sea eagles hell-bent on the Highland west have turned up all along the course of the Earn, sometimes in surprising places.

☺☺☺

The seventh fairway at St Fillans Golf Course is a long, straight, east-west par four with a drystone dyke and out-of-bounds all the way up the left, and cunningly sited groves of trees on the right to thwart your progress if you happen to slice your drive. Straightness is everything on the seventh and even then it takes two particularly good blows to reach the green. Walking off that green with a par feels like a birdie, walking off with a birdie feels like an eagle, and eagles... well, they don't come along very often on the seventh. Beyond the dyke is a field, then the abrupt, rocky north face of a miniature Schiehallion of a mountain called The Birran. Its handsome, dominating presence for an east-west traveller announces in the most unambiguous terms that you have just crossed the Highland Edge. From here to Mull, all is mountainous. Both of Scotland's eagle tribes – the golden and the white-tailed – know this better than any other creature in the land.

I have been a member at St Fillans for a dozens years or so now. It may not be the most exacting challenge in Perthshire golf (it is, after all, only a few miles over the hills to the Ryder Cup venue of Gleneagles; the eagle associations just keep piling up hereabouts), although it is deceptive and no one takes it apart, and the challenge has always been more than exacting enough for me. Its aficionados include Sandy Lyle (who lives not far away and whose sixty-six in the club's centenary year is the course record) and Ian Botham, who both know a good piece of golfing terrain when they see one. I like it for its landscape setting and for the many distractions of its wildlife when my golf is not all it might be, as well as the fact that it's a lovely place to play golf.

As for the eagles, well there are golden eagles nesting in the hills both north and south of Strathearn, which is also to say north and south of the golf course. St Fillans lies at the east end of Loch Earn and the beginning of the River Earn's journey east to meet the Tay. Those golfers who happen to know what they're looking for catch occasional glimpses of golden eagles high above both hill skylines; those who don't sometimes confuse "eagles" with buzzards (of which there are many), ospreys (an increasing presence; they fish the loch and the river and haul their catches low over the course en route to distant tree nests), and red kites that occasionally and fearlessly haunt the fairways. There are red deer and roe deer; the roe spring out from copses from time to time and the anthem of the red deer rut tumbling out of the throats of mountain corries is one of many glorious accompaniments to your golf in the autumn. There are brown hares on the course, and they too are disdainful of human presence. Wild goats eye you with an ancient stare from the hills and woods. The peregrine falcons on the mountains' lower slopes are frequent, if brief, visitors, loudly terrorising the lesser birds and delighting the birdwatchers among us. There are two species of woodpecker – the green and the great-spotted – and once I saw a grass snake sunbathing on the second fairway. So, there is a lot going on here, and for a golfer like me whose solitary aspiration is to break eighty even once a year (an aspiration mostly unrealised), there are many, many consolations to be found by breaking that most cardinal rule of golf – lifting my head.

And three times now and counting, I have been in the right place at the right time when a sea eagle has flown the entire length of the seventh fairway, and whether it is always the same one, and whether it has hares on its mind or the journey to Mull (for now the seventh fairway in my mind is also a kind of refuelling station on the eagle highway), and whether it is aware that it has recently entered the first golden eagle territory of its westward march... all these are questions to which I am still seeking answers.

Meanwhile, the thing about the seventh at St Fillans was that it added a new perspective to my awareness of the sea eagle. Despite my growing familiarity with it all along the eagle highway and as far north as Skye, as far west as Mull and Ardnamurchan, and as far south as the Mull of Kintyre, St Fillans was where it caught me unawares and put me in the shoes of other people; people for whom the sudden appearance of such a huge predator at such close quarters and in the midst of their workaday world is an extraordinary and intrusive event. The first sudden appearance of the sea eagle flying head-on and about fifty feet up above the seventh fairway was frankly startling, because it was out of context; not out of context for the eagle, but it was out of *my* eagle context. Mostly when I encounter eagles it is because I am looking for

Above: A golden eagle photographed in Harris. The wingspan of an adult golden eagle is often more than two metres.

Previous page: A golden eagle chick remains in the nest for around ten weeks before taking its first flight.

All photographs © Laurie Campbell

Below: An extremely rare sight: this golden eagle, photographed at midsummer near midnight, sleeps undisturbed on a ledge.

Above: A white-tailed sea eagle in flight.
Overleaf: Two white-tailed sea eagle chicks and (*following page*) a young fully-fledged sea eagle.

them with the specific purpose of writing them down, so I am wearing the kind of clothes that are routinely necessary for working in the eagles' environment. I am armed with binoculars, a camera, notebooks, pens, pencils and pencil sharpeners. Anticipating the appearance of eagles is my frame of mind. The closer the encounter, the better. The more graphic the birds' behaviour, the more I like it. My response to the story about the woman who was horrified at the sight of a sea eagle attacking a swan in flight was a wistful: "You saw it take on a *swan*? You lucky bugger!" I should add that I adore swans, that I have studied them and written about them more than any other creature, but I would have given a lot to witness that particular spectacle, and to weigh up the new consideration it presented that swans, too, are having to make adjustments to the march of the sea eagle across the land.

But on the seventh at St Fillans I was wearing a short-sleeved shirt, light cotton trousers, and golf shoes, and armed with nothing more than a golf bag and a dozen clubs, and my only writing materials were a three-inch pencil stub and a scorecard. So not much room for poetry.

This was the first time an eagle had found me in circumstances for which I was unprepared, and even if there was no swan or no prize goose in its sights, I concede that I did react differently. Having found me, it appeared to home in on my solitary figure on what was a particularly quiet golf course at the time. I was a creature moving across a more or less empty landscape and well lit by low evening sunshine. Eagles will often check out solitary people if they are confident in their surroundings, but a fundamental difference between our two eagle tribes is that the golden eagle is much choosier about the landscapes that inspire confidence, and for that matter, how close it will venture deliberately towards its ancient adversary, which is your species and mine. Or it could be that the sea eagle simply has a shorter memory. It certainly displays no fear or even respect in the proximity of the species that once annihilated it.

I had stopped in the middle of the fairway to watch, one hand at my brow to shield my eyes from the sun. The eagle advanced towards me in the grand hun-in-the-sun tradition of hostile fliers, and was it my imagination or did it dip in flight to improve its ever so slightly menacing aspect with the sun at its back and its wings the size of mainsails, and the whole apparition apparently over-dressed in far too many feathers? But then it shied right a few yards and watched me – one-eyed again – as it passed, steadily working its wings but flying slowly for all that. There had been a moment when its sudden and colossal other-worldliness stupefied me, and I had caught a glimpse of how easy it would be for fear, panic, even loathing to rise in the gorge of someone caught off guard by such a manifestation of nature on the doorstep or, say, low over a field of new lambs. If you were unknowing, and unaccustomed in the ways of such a bird, the almost inevitable over-reaction is not hard to understand – to reach for the phone and call the papers with your outrage.

My own second response was to revert to type. In twenty-five years of writing about nature for a living, I have learned that on the rare occasions when I feel at odds with nature it is always a worthwhile plan to take a step back and look harder at what is going on. It's easy enough once you accustom yourself to the discipline. So I had stopped walking to see if the eagle might come close. That was my second response, the come-closer invitation, because I know eagles, because

I see them often, because I admire them, because I thrill to them, because I want always to know more about them, and because for the last two years in particular they had dominated my working days and more or less filled my working hours while I tried to write them down in this book.

It was the first time I had associated the golf course itself with eagle habitat, because until recently there had been only the high-up and far-off glimpses of golden eagles to relate to hereabouts. But when a second sea eagle turned up two weeks later (or the same one returned) and a third the following year, and I found a palm-sized sea eagle pellet on that small rock that adds an edgy frisson of concern to a thinned tee shot on the par-three sixth... by then I had accepted that the green wedge of land between the young River Earn and the miniature Schiehallion had become a distinctive tract of the eagle highway.

⊙⊙⊙

The fact that the wandering sea eagle now shares the golf course's airspace (and for that matter the airspace for several miles east and west of it) with hunting and nesting ospreys has lured me into a new train of thought. The osprey is a specialist, a fish-eater to the exclusion of everything else. Its life is linked to water from the moment it spills from the egg. No water, no fish; no fish, no osprey. It is the simplest of equations. The sea eagle is a generalist with a diet that includes fish, all manner of birds, especially seabirds and wildfowl and including species as big as cormorant, gannet, geese (and apparently swans), a wide range of mammals. But its life also begins with a direct link to water, and fishing (at which it is extraordinarily adept) and all the other fruits of the sea are high priorities.

Loch Earn lies immediately to the west of St Fillans. Popular tourist legend suggests the name derives from "erne", an old Anglo-Saxon word for a sea eagle, implying a historic allegiance to the place. I am unconvinced. In the hills immediately to the west there is a watersheet called Lochan an Eireannaich, a crag called Leum an Eireannaich and a mountain whose major burn feeds Loch Earn called Craig MacRanaich, and all these are to do with *Eireann* as in Irish, rather than erne as in eagle. *Eireannaich* is an Irishman. Besides, this is a landscape named by the Gaelic language and "erne" is Anglo-Saxon. More likely by far is that the Irishman's Lochan and the Irishman's Leap (the same word – *leum* – is used for a frog; the Gaels knew what they were doing when they named things) and the mountain whose name is probably a contraction of "Crag of the Son of the Irishman" mean that Loch Earn is Loch Eireannaich – the Irishman's Loch. The Gaelic word for eagle is *iolaire*, and historically it is as likely to have referred to either species of eagle. Seton Gordon, the pure source of all subsequent twentieth and twenty-first century nature writing, noted that in some parts of the west, the hillman referred to the golden eagle as *iolaire dubh* – "black eagle" possibly because it was mostly seen against the sky as a silhouette; and the sea eagle as *iolaire buidhe* – yellow eagle, an obvious reference to that yellow billhook of a beak. The much-quoted mantra of today's publicity material for sea eagles issued by organisations such as Scottish Natural Heritage and the RSPB, that the bird is known in Gaelic as *iolaire suil na greine* – eagle with the sunlit eye – is a poetic fancy at best, and a crude PR slogan at worst.

Whatever the origins of Loch Earn's name, today it is once again a favourite haunt of ospreys, nature's rich pickings much enhanced by a trout farm. But although sea eagles drift over the golf course and turn up from time to time in the hills around nearby Glen Ample where they mix with the neighbourhood golden eagles, I have never seen them fishing the loch and have never heard of them doing so. That doesn't mean it never happens, but it is a rare event if it does.

So what I have been wondering is this: do young, reintroduced sea eagles deprived of the teaching of parent birds take longer to learn to fish than naturally raised birds? Do they not fish Loch Earn because as yet they have not learned how? And if that is true, does it not surely disadvantage them in ways we may not yet understand? And if it does, what does that say about the very core consequence of any reintroduction project based on captured nestlings – that they are compelled to make their way in the world in ignorance, unparented, unadopted, unfostered?

Is it not a circumstance that creates inevitable casualties among birds that blunder in where they are unwelcome and attempt to prey on forbidden fruit? And does that explain the poor decision-making of the bird that chose the minister's prize geese in his back garden rather than the fish-rich Firth of Tay no more than half a mile from the garden in question? I am inclined to believe that the presence of experienced parent birds would have imparted enough technique and information in even a first-year bird to prevent that kind of blunder.

In comparison, the young osprey is over the water with its parents as soon as it can fly, and trying to catch its first fish for itself before summer is over, having learned the technique, however imperfectly, from its parents' example. It's true that it must learn before its first migration at around five months old, and it must fuel that flight with the fish it catches for itself, so arguably the pressure to learn fast is more acute than it is for a young sea eagle. The young sea eagle encountering a fishing osprey for the first time cannot learn from it, for the techniques they use are utterly different, but at the very least the eagle must surely experience food for thought. What it might see is this, which is a thing I saw on a Highland Edge watersheet not too far from Loch Earn as the sea eagle flies:

A young osprey with no more than a month's flying time under its belt has been more or less stationary in the middle of my binoculars for about two minutes now. Despite its scruffy first plumage and gap-tooth wings, it has got the hang of hovering. It is, I imagine, a good osprey wind: an easy breeze out of the south-west is just enough to lean on, but not so much that the surface of the loch is overly troublesome with boat-slapping wavelets, and in the lee of low shoreline trees in this corner, fishing conditions are all but perfect.

It is noticeable, however, even to a non-fisherman like me (I like my salmon oak-smoked or in steaks, my trout oak-smoked or grilled, my herring in oatmeal, and my haddock battered), that all the three-men-in-a-boat crews of anglers who pay heftily for the privilege of catching a few of the trout hatchery's crop are elsewhere. The nearest I can see to the osprey's chosen corner is about half a mile away. I would have thought the osprey might have noticed that too, but there again it's still learning the ropes and it may well be that mastery of hovering is quite complex enough without also having to worry about where one should be hovering. One step at a time.

August is never my favourite month. It is summer in overdrive. The season is too green, the trees are too heavy, the undergrowth too thick, the light too flat, the sky too thundery, the midges too full of themselves and my blood, the mountains too busy, and the Highland roads too clogged with too many people who seem never to learn that August is when the Highlands look least like themselves. But I am pleased with the osprey's decision (if decision plays any part at all in the process, which at the moment I doubt) because it has picked my shore and its tranquil offshore waters to while away a piece of this early August late afternoon.

And I am not here by accident. This corner of this shore has its back to the sun, the shoreline willows offer cover, and the light over the water is perfect. Now, if it weren't for the midges, the flies, the clegs... but there again I have a spotted flycatcher for company, and the particularly exquisite entertainment it has offered me this last half hour also takes a prodigious toll on all of the above. Silently, I cheer it on with every strike.

Then the osprey turns up. I suppose gliding is the correct terminology for its passage into the small bay in front of me, for it involved no wingbeats, but gliding implies co-ordinated smoothness and even grace, and as yet its flight involves none of these qualities. But then it rises a dozen feet in the air, which takes it above treetop height and into the wind, where it stops, wings a-tremble, head-down, eyes boring into the dark shallows, while the wind seems intent on ruffling every undisciplined feather in its body. So almost every square inch of the bird is constantly and apparently clumsily on the move, except that at the same time its head is dead still.

It also becomes spotlit the moment it rises above the tree-tops out of shadow and into sunlight, a manoeuvre it achieves by half-closing its wings and thrusting its chest forward, like a fulmar jumaring up the thermals of a cliff-face. Not bad for just a month of flying time. But in that light, and in the binoculars, and fifty yards away, and fifty feet higher than my shore, and for as long as it holds steady on the air, every detail of the bird is at my disposal. This is birdwatching out of the top drawer.

A flying osprey tucks feet and legs neatly away under its tail. Once it begins to hover, however, feet and legs lower and dangle, and these are vivid white and thickly feathered to just below the "knee", then grey shins and feet. When ospreys attack they fall from their hovering stance and crash-dive into the water feet first, and I imagine there is no more unnerving moment in its life than the first time it does that. The osprey in my binoculars looks suitably unnerved.

Then it is gone. It has not dived, but instead it has wheeled away across the bay, effected a wide, banking turn, dropped to within ten feet of the surface, and embarked on what looks to non-osprey eyes like a bombing run. Four times in about a hundred yards its lowered feet crash into the water, but the flight's momentum is unimpaired – until the fourth one, when a wingtip also catches the surface and throws the bird sideways and chaotically into the water.

The creature that sits on the surface bears more resem-blance to the wrecked sail of an overturned dinghy than to an osprey. Its wings are laid flat and face down on the water. The bird looks around, recovers, then heaves forward and upward and is at once laboriously airborne again. A few feet clear of the water it performs the kind of mid-air shake that the adults do when they have just caught fish, discarding water in what looks like gallons.

The young osprey tries again, the same bombing run technique, with the same hapless conclusion. A minute later the bird is back in my corner of the bay, head-down and hovering. What now?

I can almost read the thought processes at work. It really doesn't fancy the dive from fifty feet. There again, the alternative strategy has achieved two soakings and no fish. A month or so from now there is the small matter of flying to the Mediterranean, which at this stage of its life, on this August afternoon, may – or may not – be an idea already implanted in its head by nature, by the fact that it is an osprey.

Suddenly the bird drops ten feet then hovers again. Then without warning it heels over on one wing, hauls in both wings to its body, pushes its feet forward and falls. The splash is enormous. When it subsides the bird is sitting there, looking round, the still centre of widening circles of troubled water, wings laid wide on the surface. Ten seconds, ten more, ten more. I feel unaccountably tense. I instruct myself to breathe out.

Then the familiar first stroke of the wings designed to heave the bird up onto its tail so that it can move forward on the water, then the first downbeat, and the next, and the bird is flying and its legs are dangling and in one of its talons there is a fish that cannot weigh much more than a quarter of a pound. The osprey is ten feet above the water again when the fish escapes and hits the surface with the kind of plop I can make with an index finger and the inside of my cheek. The osprey flies towards the shore, to a low willowy shrub where it perches, none too certainly, on the topmost edge. From there it begins to call out. Osprey voices, like golden eagle voices, are disappointing things, thin and high-pitched and just a little bit weedy. This one keeps looking at the sky as if it is beseeching an absent parent to come to the rescue, to offer advice, or preferably, fish. None comes.

For perhaps half an hour the bird sits on its willow perch without moving, other than to scratch a feather with a claw, to preen with its beak, and to look hopefully at the sky. Then it flies again, resumes its stance on the air in the quiet corner of the loch, with the sun in its eyes and the wind in its wings.

I wonder if the dive I witnessed was that bird's first, if the tiddler that got away was the first fish it had ever caught. From here on in, it is mostly all just refinement of technique and the lessons of experience. Hunger will take care of the rest. This bird will catch fish because it is an osprey and ospreys catch fish. Given a fair wind, this gawky juvenile will be back here two or three summers hence, and it will help to swell the steadily growing breeding population for which this loch is a midsummer lifeline.

The people who run the trout fishery here like having the ospreys around, because the fishermen like to watch them, and to swear at them from time to time when one interrupts a barren hour of human fishing by diving in between two boats and snatching a two-and-a-half pounder from under their noses. Back in the bar in the nearby hotel, the traditional fishermen's stories are garnished with exploits of the ospreys they see, and always the osprey in the story carries away a bigger fish than the one in the previous story. But the one that chose the quiet south-west corner that quiet August afternoon is spared their fantasies. And this was its true story.

But surely, and surely sooner rather than later (if indeed they have not already started), the fisherman are going to start telling sea eagle stories, and their arms are not wide enough – not half wide enough – to do justice to the size of the one that got away with the fish in its talons, the fish that it caught without even getting its feet wet.

☉☉☉

Mike Holliday is both a gamekeeper and a good friend, a combination that is something of a rarity in my life. He is also as acute an observer of wildlife as I have ever met, and he will talk golden eagles all night with passion and frank admiration. He lives and works in the hills south of Loch Earn, whose main thoroughfare is Glen Ample, a north-south glen whose two burns feed Loch Earn in the north and Loch Lubnaig in the south. It is one of those edge-of-the-Highlands glens where golden eagle territories are perhaps not so clearly defined as in some heartland and west Highland strongholds. These are also particularly busy hills for walkers and climbers, and include two of the Munros closest to Scotland's central belt population. Human disturbance is one of the most unsettling factors to affect golden eagle breeding and territorial behaviour. Glen Ample is also something of a thoroughfare for migrating birds in spring and autumn, one which, in conjunction with Glen Ogle to the north of Loch Earn, lifts many thousands of birds a year into and out of the Highlands (see Chapter 7). And although Glen Ample does not hold any golden eagle eyries, its airspace is contested by breeding and non-breeding birds for its seductive combination of prey species and hunting winds and thermals that are forever at work in the glen's repertoire of corries, ridges, crags, buttresses and high moors. None of this is lost on passing sea eagles, which turn up with increasing regularity. The very presence of golden eagles will lure them in, as it seems to do throughout the natural range of golden eagles, unquestionably a factor that lures west-making sea eagles ever deeper into the Highland west. Time will tell if the process works in reverse, or if the west-coast-of-Norway DNA in all Scotland's sea eagles is too irresistible a force to permit an east coast colony to thrive.

I was curious to know how the sea eagles' increasing presence was affecting things among these frontier golden eagles, and few people know them better than Mike Holliday. So I went to see him, to talk eagles. The thrust of his opinion was this:

So far, the sea eagles are making no difference to the golden eagles. Bear in mind that the sea eagles we have been seeing are all young birds in a strange landscape, and the golden eagles are either breeding birds on their hunting territory or their offspring, and they are part of a lineage with an unbroken historic attachment to and intimate knowledge of the landscape. I thought they might have competed with golden eagles for red deer grallochs that stalkers leave out on the hill, and which are an undeniably valuable source of food, not just for eagles but for all manner of nature's Highlanders, from foxes and ravens to red kites and buzzards. But apparently not. Firstly a golden eagle on a carcase (and a gralloch is a discarded piece of carcase) is a difficult creature to shift in a country that has long since done away with its wolves and its bears. Secondly, although a sea eagle is much larger, a golden eagle is much the better flier, and the odds-on favourite in a physical joust. Besides, the sea eagles are not competing for nest sites, not here, not yet, and perhaps never. For me as a nature writer and for Mike as a keeper and eagle addict, one of the most alluring aspects of the situations we find ourselves in is that thanks to the sea eagles, nature is re-writing the rules before our very eyes. People are behind the reintroduction project of course, but only nature controls its outcomes. One thing we were both sure of – the golden eagle can out-fly the sea eagle every time. And for reasons that are not particularly easy to articulate, it is an outcome that pleases us both.

So the conversation shifted to the glories of golden eagle flight. I said that I thought there was nothing – nothing at all – in terms of bird flight that a golden eagle couldn't do, including flying backwards; that it was the only bird I had ever seen fly backwards deliberately and in complete control of what it was doing. Mike nodded in agreement as I began to tell the following story, a nod so emphatic that I guessed at once he was waiting his turn and then would top my story with one of his own. He was. He did.

But first, my story was this. I was in the eagle glen a few years ago and sitting on my accustomed perch by the big rock. The resident male crossed the east ridge and started to head over the glen towards the eyrie but travelling about a thousand feet higher. Halfway across he half-folded his wings and dived at a steep angle towards the rock wall that accommodates the eyrie. He was about a hundred feet *below* the eyrie itself before he opened his wings and swept up the rock face, rose to perhaps ten feet *above* the eyrie where his mate was now standing on its outer rim, became motionless with wings wide open, then started to ease *backwards* on the wind or using a kink in a thermal – demonstrably I have no idea! – until he was directly over the eyrie and set himself down with the delicacy of thistledown. I felt like applauding.

Mike smiled, then began to unfold his account of a most extraordinary day among eagles.

"It was one of those days when you get the full moon in a clear blue sky, and it was also the day that I missed out on the opportunity of the best golden eagle photograph ever. The moon had just risen above a hill shoulder and I had been watching an eagle flying slowly above the ridge. Suddenly it flew right across the face of the full moon. There was one moment when the outer edges of its wings were just contained within the outer edge of the moon. If I'd had a decent camera with me... one good shot of that and I could have retired for life."

So I was still grappling with the image he had just laid before me when he said:

"I thought that was it for the day, that nothing could top that moment, but later the same day I saw the male eagle pull his wings in and go into a terrific stoop. It's always a thrill to see that. But as he came out of his dive and started to rise again I expected him to start working his wings, but instead he hugged them against his body, so he was climbing only by momentum, so he got slower and slower as he rose until he stalled. Then – like your eagle – he started to drift backwards on the air until he was more or less back to where he had begun, and then he dived again, and then he did the same thing at the bottom of the dive... climbed without using his wings, stalled, drifted backwards, and then he did it all again."

By this time, I was shaking my head and probably grinning like an idiot. But...

"And then – the second eagle. She flew up to join him, and then they were both doing it!"

At this point, Mike began demonstrating with his hands held out in front of his face, each hand representing an eagle, pulling the back of one hand towards him to represent one eagle flying backwards, then showing how the two eagles choreographed the performance with overlapping vertical circles, the male high when the female was low, the female climbing as the male dived and drifting backwards as he climbed... and it went on and on.

"What do you think was going on? Were they just enjoying themselves?"

"Just that," Mike said. "They were doing it because they *can*."

Our talk wandered back to sea eagles. I told him I was gathering stories from a few friends about the first time they saw a sea eagle, and I told him Ann's story from Tentsmuir, then asked him about his, suspecting it would be a bit different from the others. It was.

"It's hard to forget it," he said. "It was in the glen and it was sitting on the edge of a crag and looking like a gargoyle. At first I wasn't sure what it was, except that it was big and I didn't think it was a golden eagle. Then I saw that huge beak. Then I saw that it had a transmitter with a small aerial on its back and it was being hassled by a guy in a microlight with a receiver."

"*What?*"

"The guy was tracking it illegally. You could see the dish in the front of the microlight. I phoned up to report it but they knew about him already: my description of the microlight matched.

"But the closest view was when we were burning the heather and two sea eagles came straight through the smoke, side by side. That was impressive. They came right over and had a good look at us, and vanished."

Coughing, presumably.

Chapter 7

THE CENTRE

THE COTTAGE STILL CROUCHES a little below the main road through Glen Dochart. It has been spruced up since my day, when it leaked weather through walls, windows, doors and roof, and field mice nested in the sofa, among other places. Glen Dochart is an ancient route through the heart of Scotland, and something of a bridge between southern and northern Highlands and for that matter, eastern and western Highlands, but without ever seeming to encompass any of the qualities of any of them in particular. From the early spring of 1998 until the autumn of 2003 I lived here. It was like living in a temperate rainforest without the forest, which has mostly been removed.

More days than not, I saw golden eagles from the back garden – "garden" being a euphemism for a fenced-off piece of hillside – but the eagles and the situation more than compensated for the shortcomings of the cottage. Whatever the shortcomings of my life at the time (and these comfortably outweighed the shortcomings of the cottage), I worked well here. In five years I re-wrote and revised a new edition of my Cairngorms book, *A High and Lonely Place*, to mark the tenth anniversary of the original; I wrote my two novels, *The Mountain of Light* and *The Goalie*; and a collection of nature essays, *Something Out There* (all published by Keith Whittles). I also spent almost a month in Alaska to make two radio programmes for the BBC Natural History Unit. I have rarely been so creatively productive.

On good days I used to haul a small kitchen table out into the garden to write in the sun. I had buzzards for near neighbours, four of them in the sky at any one time was routine, but I got to recognise that when all four leaped from the two small spruce plantations behind the house at the same time and began crying incessantly, it usually meant an eagle was in the offing, cruising across the lower rabbit-thronged hillside. The tract of rock and bog and rough grazing that climbed from the back garden to the mountains seemed to mark a notional northern frontier for two pairs of golden eagles, or perhaps it was just that they shied away from the main road and if they did cross it, they liked a lot of height between them and its human traffic, for I never saw them cross it once.

Four buzzards acting on the same impulse and in common cause was a conspicuous enough event to lift my head from the work. I scanned the sky first, but it was empty, then the nearer low ground and there I found the eagle, low and slow and with a not-quite-dead rabbit still twitching in its talons. The eagle was a young male, his white wing-and-tail patches showing his immaturity, and he had a peculiar relationship with the resident pair in the eagle glen beyond the watershed. His ambitions centred on the huge dark female (the pale female's predecessor in the eagle glen) but her mate saw to it that he was not allowed to linger near her, not that that stopped him from trying. So he would cross the watershed and head for the rabbit warren where he would weave sound-lessly through the piercing, four-fold hostility of the buz-zards, and with his talons lowered, choose a rabbit from the hundreds on offer at any one moment, then turn and labour back to the watershed and (if the coast was clear), deliver the rabbit to the eyrie. It was extraordinary behaviour, but he did it again and again throughout that long nesting season.

Mostly, over the decades of watching the eagle glen, I have climbed to it and to the watershed from the south. The Glen Dochart years altered my perspective of the place, not least because my routine walks from the back door took me immediately into the eagles' hunting territory, but also into the territory of the neighbouring pair. The territories seemed to fluctuate (as much with wind direction as anything), but the burn that fell from the hills to my garden was a rough demarcation line: a mile east or west of it, and I was sure of the eagles I would see.

There was an afternoon of mid-July, by which time the day's writing was bogged down and going nowhere. (Ernest Hemingway insisted that you should quit your shift while the work was still going well so that you would pick up the good work when you sat down again the next day. Easy for you, Hem, but I never had that kind of courage. I wrung every drop of nectar from the good days in case there was no more nectar.) It seemed to me back then that there was a species of rain unique to Glen Dochart as merciless as loneli-ness. It shuttered the landscape, stilled and drenched the air, silenced birds, and seeped greyly into a hungry human heart. Sometimes the writing mind is a two-edged sword that feeds on the solitary nature of the process out of necessity and which in turn demands that the writer's resources are robust. That

July afternoon, both edges were blunt. Then I looked up and saw that the room had lightened. I took my unworthy mood to the back door, swathed it in waterproofs and boots, threw some food in a small pack and closed the door behind me.

The air smelled of the woods and the rain. Beyond the woods it smelled of bog myrtle, that sublime reviver. The rain was faltering at last and I peered up through its fraying curtain towards a clearing hill shoulder from where the shouts of crows drifted down. There were three of them and they were mobbing an eagle that was perched on a rock. I had been seeing an eagle above that skyline regularly over several weeks, and assumed it would be the young male drifting in and out of the eagle glen territory into the vacant one nearby (a mysterious vacancy, a sudden and unexplained gap in a long history of continuous occupation; what changed?).

An hour up and out from the cottage I reached the skyline, and sat for a breather on the sodden mountain among thinning shreds of cloud. It was quiet and still, and I felt my mood lifting, the day breasting a watershed of the mind. Then a voice sounded from a nearby rock, a voice like the shyest of penny whistlers, as if the player had changed his mind about playing at all after the first note had sounded, so that the note had a tiny upward inflexion as the fingers lifted off. It was the voice of the mountain's softest alarm call and the signature of a golden plover, gold backed and crowned, black faced and fronted, and between gold and black an elegantly curved and carefully worn scarf of white. The bird stood blatantly on its rock, a confident profile, the short, dark bill opening and closing again and again to play the shy, one-note whistle, relentless as cuckoos. It is the easiest of bird calls to imitate and soon we were penny-whistling to each other like two old cronies at a ceilidh. But with my head turned towards the plover and my preoccupation with its music, I missed something that was unfolding at my back, until that inexplicable sensitivity to something unseen and unheard but rather *felt,* like a warm breeze on my cheek, demanded my attention and I turned away to look up the ridge.

As I turned, a piece of the ridge detached itself and became mobile, became low-flying fluency, became eagle, not a quarter of a mile away and closing. The thinning cloud had revealed my presence as an unfamiliar shape on the eagle's landscape, and this being the perfect example of a golden eagle confident in its landscape, it chose to inspect me. It flew past about fifty yards away and slightly below the level of the rock where I perched, looking at me side-headed – that one-eyed glare again – with fire in its eye. I raised a hand, hoping the bird would read the submissiveness of its gesture, but as likely as not it saw only my intrusion.

And this was not the young bird from the far glen but a mature male with a pale head but not a shred of white in his plumage. Then again something turned me round to face up the ridge once more, and a second eagle rose from the ground not a hundred yards away. How had I missed that? How long had it been there? In its talons as it contoured away round the hill was the slumped shape of a small bird. In the few seconds when I was able to bring the glasses to bear on it, I recognised the pale gold of a female golden plover. So I turned back to the penny-whistler's rock: gone without trace. His is one of the lesser shapes of the hill that immobilises and trusts in stillness when the eagle shadow crosses.

Questions:

Was this a new pair of eagles?

Was there an eyrie much nearer to the cottage than I had imagined? All this was at the "wrong" end of the glen from the traditional eyries, yet if this was a new pair on an old territory, was it so unthinkable that they would work from a new eyrie site?

If there was no new eyrie, why was the second eagle carrying prey away round the hill in the "wrong" direction?

More circumstantial evidence: the second eagle suddenly reappeared crossing the skyline without its prey, so the prey had been dropped off nearby, or at least dropped.

More questions:

A new eyrie just round that hill shoulder?

Just beyond the skyline?

A territory no one else knows about (a wholly unworthy thrill smote me at that prospect)?

Then the two birds were together, flying on parallel courses far out across the glen, cruising at two thousand feet. I went in search of answers. Four hours later, four hours of knee-wrenching, foot-sliding, muscle-wearying contours and crags and spaces between crags, I arrived back at the same rock to be met by the same golden plover on its rock. I had answered none of my questions, but I had satisfied myself once again about the power of the high and lonely places to pacify a bruised human spirit, the solace that I have always derived from the terrain of eagles.

I pulled my small and well-worn whisky flask from my pack, raised it towards the hills that accommodated the timeless coexistence of eagle and plover: "Until the next time!"

Until I climb again, until I trespass again on the four-season realm of eagles, the spring and summer realm of the penny whistler that may or may not have still had a mate, the lifelong realm of the nature writer with a wilderness thirst to slake and a headful of questions that demand answers, so that they can become words on the page and improve his understanding of his portion of the wild world.

⊙⊙⊙

Two years after I left Glen Dochart and moved south over the hill to Balquhidder, the east coast sea eagle project began. Seven or eight years after I left, I had begun to piece together the notion of the eagle highway as more and more sightings became common knowledge at many points along the route I had been guessing about – the Tay estuary to Mull. By the time I had established for myself that the route diverged where the Earn joined the Tay, I had a good idea from the nature of reported sightings further west that the two divergent branches converged again on a single centrepiece – Glen Dochart. Sea eagles that follow the Earn will gravitate naturally towards the west end of Loch Earn, for there, right above the village of Lochearnhead, is a crossroads as old as the very shape of the landscape for travelling birds of many tribes. The tendency of the land hereabouts to open up in east-west glens is burst apart by a great north-south intrusion. Northwards then westwards, Glen Ogle carries the high road to Glen Dochart; southwards, a two-fold trough carries both Glen Ample and the low road to Callander, Stirling, and the Lowlands; Strathearn itself opens up the east. Spring after spring for most of the ten thousand or so springs since the last ice age finished reshaping the land, hordes of geese have hesitated at that crossroads high above Lochearnhead, circling and climbing, circling and climbing, until at last they achieve the necessary height and visibility that shows the way north then north-west, next stop Iceland, Greenland, or wherever. Autumn after autumn their migrations split apart here – south or east. They are the most visible manifestation of the role that crossroads plays in nature's annual rituals, and their behaviour pattern is mimicked by countless other bird travellers. The west-making sea eagles, even in their infant years, cannot fail to recognise the landscape significance of that crossroads. Some will baulk at that wall of hills with its single narrow escape route to the north and west, and these will turn south instead, perhaps to follow Glen Ample and cross the high moors east towards Argaty, near Doune, where large numbers of red kites routinely swarm around a feeding station to the delight of human visitors in a nearby bird hide. Sea eagles are regular visitors there. Others have dallied among the lochs and rivers of the Trossachs, but as yet they have never lingered long. But mostly, the birds' pursuit of a sunset-facing, island-strewn west coast, fuelled by thousands of years of Nordic inheritance, will point them towards Glen Ogle, and lift them over its high pass and down into the gathering centre of all that land where the eagle highway unifies again and strengthens into the way west – into Glen Dochart.

For birds that travelled by way of the Tay, the journey has perhaps been simpler. The river has led them naturally to Loch Tay where the mountains on the north shore and lower hills of the south shore channel the way west. And it is at Killin at the west end of the loch that the River Dochart itself swerves gracefully into Loch Tay, carrying with it water from the ultimate river source in all Scotland, from the Tay's furthest tributary, a tiny lochan high on a flank of Ben Lui, not thirty miles from the Atlantic coast, and from where the mountains of Mull must surely look like nothing so much as a glimpse of Valhalla to a Norwegian-born sea eagle. At Lix Toll, two miles west of Killin and five miles north of Lochearnhead, the two divergent streams of the eagle highway reunite in a single confident westward thrust of momentum. A mile west along Glen Dochart's flat-bottomed glacial valley, then, the cottage still crouches, and there are new eagles to add to the endless march of travellers that pass a few yards from the front door.

⊙⊙⊙

I was double-crossed by the month of March on the ninth. For its first eight days that least predictable of Highland months had blithely picked up and run with what had stretched out of late February into a two-week spell blessed with portents of spring: dawn birdsong, drumming woodpeckers, nest-building ravens, the daffodils' silent crescendo of goldening trumpets along the bank of the burn, the stupendous sky dance of golden eagles a thousand feet higher. But all that was gagged, hobbled, and refrigerated by a snow-burdened north-easterly that would do its worst for the rest of the month and deep into April. On the ninth, the day when the Judas wind shuddered down the east of the land, I hastened west just beyond the reach of its blizzards (although its Arctic breath was on the air) to the deep green sanctuary of a Scots pine wood on the shore of Loch Tulla, six hundred feet up in the east-facing folded lap of the mountains of Glenorchy and Inishail.

Glen Dochart, and to a certain extent Strath Fillan beyond, is a kind of decompression chamber that sheds lesser notions of Highland landscape and equips the traveller's mind to accommodate the different intensity of that landscape that lies beyond, whether the journey is further west or north. It is not exactly an original concept. I encountered something like it in the Ardnamurchan poet Alasdair Maclean's book *Night Falls on Ardnamurchan* (Penguin, 1984), in which he described the Corran Ferry as "a kind of mobile decompression chamber where various kinds of pollution were drained from the blood and I was fitted to breathe pure air again". I was very taken with the notion that it was helpful to adjust from the uncivilised influences of the mainland at your back to the island-like peninsula of Ardnamurchan, and that the ferry provided the crucial respite.

Likewise, the southernmost Highlands around Balquhidder have a beguiling landscape softness that dispels in Glen Dochart, but it is only once you are beyond the glen and through the village of Tyndrum that the road kicks northwards in a great climbing loop, and from that moment on, the mountains close in and crowd round and the Highlands become a different creature. Even the vast open spaces of Rannoch Moor feel more like a mountain plateau than a huge bog, and as soon as you cross it, there is the small matter of Glencoe to deal with. But it is the towering, unbroken sweep of Beinn Dorain's south-facing profile that offers the first of many mountain milestones that will stop the traveller in his tracks and drop his jaw once he emerges beyond Glen Dochart. Beinn Dorain is the Mountain of the Otter, apparently, although no one you might ask seems to know why; nor why for that matter, given that its name is a phonetic corruption, it should derive from *dobhrain* (an otter) rather than *dorainn* (pain or torment). But then the Gaels' naming of the landscape was based either on local tradition or the way that the landscape looked from where they lived, which was in communities of which most have vanished at the insistence of sundry oppressors over many centuries. Inconsistent spelling, anglicisation that was haphazard and as careless as it was uncaring, and just the passage of time that has carried off with it the sources of so many names... all that has mangled many a poetic or prosaic christening into meaningless hybrids. The land shrugs its indifference to all that. It has no opinion on what we call it, but our understanding of its history is so much poorer because time has been so careless with it.

Nevertheless, many an eagle still nests or likes to perch on, or identifies its territory with conspicuous rocks called *Creag na h-Iolaire*. And many a sea eagle, whether embedded in the resurgent strongholds of the island west or wandering that way along the highway from the east, will recognise those rocks for what they are, and be intrigued by them because in eagle eyes they are the most coveted of places, as they were before the landscape was named. Once an eagle rock, always an eagle rock. Besides, sea eagles have been here before.

Two distinguished eagle authorities two or three generations apart – John Love and Seton Gordon – have identified historic sea eagle eyries hereabouts on islands in lochs. Loch Tulla is one of these. John Love's name is synonymous with the sea eagles of Scotland because he was the Nature Conservancy Council's choice to lead the reintroduction project on Rum, and he has written the bird's Scottish history. He identified over a hundred historic nest sites in Scotland, mostly in the Hebrides, and fifty more in Ireland. Of that part of Argyll that conducts my notional eagle highway westwards to the sea, he wrote that they "frequented the mainland hills of Creran and Etive" and that further inland they nested in a small tree on a tiny island at Loch Tulla. When I read that comment it rang a bell and put Seton Gordon in my mind, for I was sure he had written something similar. I trawled through a few of his books until I found what I was looking for.

So I invited myself on a fool's errand on this short diversion to see if I could track the spoor of the sea eagle on the edge of Rannoch Moor more than a hundred years after it had vacated the premises. And at the back of my mind was the straw I had begun to clutch: once an eagle rock, always an eagle rock. Or in this case, eagle tree. Besides, long experience of this kind of escapade has taught me that putting in the hours out in the landscape I'm writing about always pays off; sooner or later something or other turns up that I would not have known about if I hadn't gone. So I turned off the main road beneath Beinn Dorain at Bridge of Orchy for the single-track road to Loch Tulla, where the old pinewood relic is a particular favourite of mine.

Alasdair Maclean's decompression chamber proposition actually works. Or maybe I only think it works because I found the idea so engaging in the first place, and having adopted it to underpin an idea in my own mind, I wanted to be convinced by it. But stepping from the car into boots, jacket, gloves and pinewood felt like being reinvented as a denizen of a more rarefied world where nature is a more demanding presence. These are harder-edged Highlands, cold, uncompromising and austerely beautiful. From now on, all the way to the north coast if you care for such a journey, these qualities only intensify.

And there, after only a few minutes' walk along the loch's pinewood shore, is the tiny island that once accommodated a sea eagle nest. It is not hard to identify. It is the only island in the loch. And it still has trees of a kind. The stony ground and the onslaughts of pretty well every wind that ever blew have restricted and twisted the ambitions of four larch trees, which could be of almost any age at all. The solitary pine that keeps them company is dead straight and slim and surely a young tree. The piled stones that showed above the waterline made me wonder if it was a crannog. The map thinks it's an island and gives it a name – Eilean an Stalcair, the Stalker's Island. I wonder if it is the smallest island in the land that has a name. And once upon a time, it had a sea eagle's nest, a thing of such monstrous dimensions that it must have looked as if it might destabilise and capsize the whole island. Any day now, if the eagle highway works in both directions, such a nest might transform the island's modest profile again. The Argyll Bird Club's outstanding book *Birds of Argyll* (2007) pinpoints the presence of the bird not just on the islands of Mull, Jura, Islay, Lismore and Kerrera, "but also, less frequently, from more land-locked water bodies, for example Loch Awe... Birds of up to two or three years old can be particularly itinerant; for example, within two years of fledging on Mull, one was seen on Jura, Islay, coastal Mid-Argyll, inland North Argyll and Morvern." Inland North Argyll means places like Loch Tulla and Loch Ba. The grapevine that whispers through these places has already brought word of that mightiest of bird silhouettes low over the waters of Loch Tulla in the spring of 2010. It did not linger but it was there, and now of course it knows the way.

So birds from the west, and especially the breeding stronghold of Mull, are spreading east and inland as well as up and down the Atlantic seaboard, and as they do so they are encountering historic territories of their species, they are encountering sea eagles making for the west coast from the east, and they are encountering golden eagles, and all of these are incentives to linger in places like Loch Tulla and Loch Ba. And given that we already know they have been seen on Loch Awe (a short flight the length of Glen Orchy south-west from Loch Tulla) and Loch Etive (a flight of about the same distance to the west from Luch Tulla), it is a brave man who would bet against the sight and sound of sea eagles nesting once again (and with delicious irony) on the Stalker's Island.

I settled into the lee of a generous Scots pine trunk and focussed the binoculars on the island. The biggest of the larches appears to have broken about two thirds of the way up the trunk, and the break appears from the distance of the shore as a dark, squat, vertical shape that reminded me of nothing so much as that unprepossessing, grey-brown, shape-less "embellishment" I first saw on a rock on Mingulay the day that sea eagles first entered my life.

It couldn't be?

Could it?

It could not possibly be that at the very moment I had turned up with binoculars and a clutched straw, and after an interval of more than a hundred years, that my visit coincided with a sea eagle visit, that it was perching on a tree on the very scrap of island that sea eagles once called home.

Could it?

I stared and stared at that dark shape, willing it to move, to throw wings like a crumpled parachute above its head and metamorphose out across the loch on articulate wings. How I stared and how I willed. Two hours later, it had declined to become animated, leaving me to conclude reluctantly that it was a piece of broken larch tree. But in those intervening hours I had discovered what was unmistakably an osprey nest in the top of a dead pine, and intact enough to convince me that it had been in use the previous year. Strange how ospreys keep cropping up on sea eagle waters, or perhaps it is the other way round. Records suggest this might be something of a new frontier in Argyll's slowly re-establishing osprey population although it has already returned to historic haunts on Loch Awe, and passing birds have paused here on October migration flights south. I chose to treat the discovery as a good omen. If ospreys can find their way here so can sea eagles, for the loch is ideal for them both and for the same reasons.

A month later, I was back, and now I was looking for eagle and osprey, and in the meantime it had occurred to me that just as wandering sea eagles seem to be attracted towards the company of golden eagles, they might also be attracted towards the company of ospreys, especially nesting ospreys. My plan was to look at the island in better light, and follow up another whisper about a sea eagle nearby, but it was thwarted by the discovery of an osprey, newly arrived and very conspicuously perched on the rim of last year's nest. I couldn't watch the island without also disturbing the ospreys, and there are laws about these things. So I drove on north for a few minutes and a handful of miles to Loch Ba.

Seton Gordon died at the age of ninety in 1978 and just three years after John Love's project began on Rum. What I had been looking for in his books was an intriguing story about a sea eagle nest on a birch tree on an island on Loch Ba. I thought at first it might have been the same island nest, but both Seton Gordon and John Love are specific enough in the details to be certain – two lochs five miles apart, two different islands, two different nests on two different trees. What we don't know is whether or not it was two different eagle pairs. Seton Gordon's adult life had been largely devoted to an exploration of the Highlands, its landscape, wildlife, people, music (he was a piper and piping judge) and its language. "A man may be said to live on after his death through his books," wrote John Muir, and that was as true of Seton Gordon as it was of Muir himself. He had the ear of keepers and landowners, and some very useful friends in high places besides (his telescope was a gift from the Prince of Wales) and he used their knowledge and their stories to amplify his own relentless and skilful fieldwork. Eagles held a particular fascination for him, and my own writing has known no higher honour than to be asked by publisher Keith Whittles in 2003 to write an introduction to a new edition of Seton Gordon's 1927 book *Days with the Golden Eagle*. There had been nothing like it when it was first published, and it has not been bettered since. I cannot watch the flight of an eagle then try and write it down without being aware of his long, long shadow. My introduction to that new edition of 2003 began:

The debt I owe to Seton Gordon is not easily measured. It is quite possible that without him there would have been no modern nature writing tradition in Scotland at all, in which case, where would that have left those of us who came after, grateful for his footprints in the snow of our landscapes? The Victorian era into which he was born was characterised in Highland Scotland by a loathsome attitude towards wildlife in general and birds of prey in particular. Yet miraculously he emerged from it looking at golden eagles through camera and telescope rather than along the barrel of a shotgun, and with a personal crusade to communi-cate to others how an intimate acquaintance with wildest nature could be both a joy in itself and a great enrichment to the human mind. In the early years of the twentieth century this was new and daring work, and it must have won him as many enemies as admirers.

It also became his life's work. There are those who still speak with something approaching reverence of hearing him lecture inspirationally not long before his death at the age of 90...

His second book about eagles (although they cropped up in almost all of his books) was *The Golden Eagle*, published in 1955. It is in that book that he relates the following story about Loch Ba:

In the posthumous edition of his book The Moor and the Loch, published in 1888, John Colquhoun of Luss tells us that, in the Blackmount Forest[2], he was able to see, from the shore of Loch Ba, both golden eagle and sea eagle above their nesting territo-ries. The golden eagle nested on a rock west of the loch, the sea eagle on the large island on Loch Ba. The golden eagle hunted on the high hills, the sea eagle over the low, boggy ground. Their hunting grounds being distinct, the two birds rarely met, but there were occasional clashes, in which the golden eagle always got the better of the sea eagle. Colquhoun lived in Victorian

---

[2] The use of "forest" in this context refers to a "deer forest", a bizarre Victorian phrase to describe estate land where trees had been removed to make it easier to shoot deer.

days, when a bird watcher did his best to secure a rare bird as a trophy: he attempted to shoot the sea eagles of Loch Ba, but was apparently unsuccessful. His story is valuable, because it shows that the sea eagle bred inland as well as on the coast. The Loch Ba eyrie was on a birch tree. As Colquhoun was being rowed out to the island, the sea eagle left her eyrie and perched on a neighbouring tree, "her white tail shining like the silver moon".

The hunter landed, his companion made a "hide" for him (surely one of the first "hides" on record), a small aperture was made for his gun barrel, and for six hours he remained in his "hide", his gun covering the eyrie. During this long watch the sea eagles usually floated at an immense height, but sometimes dived to the neighbourhood of the eyrie, beating their great wings which, he tells us, made a hoarse, growling noise like the paddles of a steamer heard at a distance on a calm day. On one occasion the eagle did indeed return to her eyrie, yet Colquhoun did not fire, and the reader has the feeling that he was half-reluctant to destroy her. At all events, after this long watch in the "hide" he did not return until the following year, and we are not told what happened then.

Seton Gordon's story makes intriguing reading for a twenty-first century eagle-watcher. Many of those who object to the whole sea eagle reintroduction project cite disturbance to golden eagles up to and including taking over their nest sites, while cheerfully ignoring the essential truth that for ten thousand years since the Ice Age, golden eagle and sea eagle thrived side by side without any need for human intervention. Human intervention in the "management" of wildlife behaviour is the point at which things usually go wrong. John Colquhoun was writing at a time when overbearing management of all manner of bird and mammal predators had rendered many species extinct and brought others to within sight of extinction. Seton Gordon's observation that Colquhoun was "half-reluctant to destroy her" is surely true: one of the reasons why the sea eagle succumbed to extinction while the golden eagle survived in the same landscape and exposed to the same attitudes and the same guns, is that the sea eagle was a much easier target, much larger and slower off the mark, and often given to nesting at ground level.

Loch Ba may be only a handful of miles from Loch Tulla, but its character is utterly different. If the decompression chamber of Glen Dochart prepares the traveller for the kind of new intensity in the landscape you can feel at Loch Tulla, Loch Ba evolves that transition further, and into something more wild, less controlled, less controllable for that matter. It is a singularly primitive place. It lies at an altitude of a thousand feet and is wide open to the tongue-lashings of every wind that was ever brewed in the imponderable space that is Rannoch Moor. It's a loch that is hard to love.

Early April, and Loch Ba was still largely ice, an eerie pale blue-green-grey shade that was oddly compelling, a thing of glaciers. And for all that, it was surrounded on three sides that early spring by head-turning mountains dazzling in sunlight and deep snow, the loch was the most hypnotic facet of the landscape. In this mood, nowhere in Scotland (apart from the Cairngorms plateau, which is three thousand feet higher) looks and feels as Arctic as this. The transformation from Loch Tulla is astounding. The psychological impact on the traveller – that you have climbed to an Arctic realm – is supported by geological time. Loch Ba is where that inland-sea-like presence of Rannoch Moor washes up against the mountains of the Blackmount. Rannoch is the architect of all this, a one-time ice cap such as you can find today about six thousand feet up in the heart of Iceland. Its first consequence was to grind down and reshape every mountain in sight, while its glaciers whittled away the land into countless glens for many miles in every direction. When all that was done, the legacy of its passing was the greatest of all our native woods that swarmed across a strewment of rock and lochan and bog all the way to far Schiehallion. We – your species and mine – took care of the trees, though their bleached bones are everywhere in the peat. What we now call Rannoch Moor is the corpse of the Great Wood of Rannoch. These birch-and-pine-clad islands and outcrops demonstrate unambiguously that with human will and the reinstatement of something like natural forces (wolves, for example, so that the red deer hordes would remember how to behave like wild deer again), a Great Wood could flourish here again, and all nature – and all of us – would have cause to celebrate. This is a place of huge gestures. White-tailed eagles here? I should think so!

⊙⊙⊙

The birches on "the large island" in Loch Ba were in their weariest end-of-winter guise, and looked deceptively pale and frail, and tawny in the sunlight, but touched with that purplish winter mist that seems to cling to all birchwoods. Most of the trees look young and insubstantial but there is one hoary old relic at the heart of the island copse. I subtracted in my head: 2013 minus 1888 (when Colquhoun wrote his book) – one hundred and twenty-five years, which is much too old for a birch to survive in the best of conditions, but this is perhaps the grandchild of the sea eagle's tree. But you need birches in a land as unforgiving as this one. Hugh Johnson, writing in his wondrous landmark of a book *Trees* (Mitchell Beazley 1973, Octopus 2010), describes birch as one of the hardiest of broadleaves, able to thrive even in the Arctic conditions of Greenland and Iceland, "...essentially nature's stopgaps, quick-growing and short-lived, graceful rather than dignified. It is their nature to lean slightly from the true; their strength is in their wayward femininity..."

"Nature's stopgaps" is perfect for describing these birchy clusters in this high latitude doing its passable impersonation of the Arctic on what Hugh Johnson would call "bare, often starved land" with its "extremes of drought or damp" — more damp than drought to be sure, but I walked across Rannoch Moor one long hot June day and night a few years ago and never saw anything in all Scotland so transformed by drought. The wolves would be the agents of more birches by transforming the behaviour of the grazing hordes, and sea eagles would be just one of the countless tribes to benefit. They would find Rannoch's spaciousness to their liking too, and now that the osprey is already back at Loch Tulla, it is a matter of time before they drift this way too. If it works for ospreys, it will work for sea eagles.

As for Seton Gordon's remark about Colquhoun seeing the golden eagle here, and that its nest was on a rock west of the loch, the "rock" in question is not hard to find today if you know what you are looking for, but on that icy early April day, I decided against the likelihood of seeing one. By April the eagle is usually on the nest, but only the south face was free of snow and golden eagles rarely use those. The rest of the "rock" was under deep snow and that is often enough to dissuade golden eagles from using their preferred eyrie site, or even to persuade them to abandon nesting for the year altogether. There is also the small matter of the disturbance offered up by the proximity of the West Highland Way with its footsore legions of gaudily clad walkers bowed down by packs the size of sea eagles and mostly looking at their feet.

Loch Tulla and Loch Ba require short diversions from the eagle highway in my mind, but diversions short and long are in the nature of sea eagles, and if the highway really does establish itself as a permanent trade route, then birds from east and west will find their way to the historic heartlands of their tribe simply because they are sea eagles and they will acquire the knowledge that leads the way.

Chapter Eight

THE HIGHLAND WEST

The river roars with the voice of an angry ghost in a long-severed Celtic head. There are places that carry their own chill on the brightest day, where, when rain darkens the hills and the mists hang low, one's eye half catches stealthy movement, one's ears hear a soft footfall. Go, if you dare, to Glen Orchy when the spates are out; stand back, for any sake, from the banks, for this is a greedy river that takes a life as readily as it tears away a rock or a tree. Stand back, the ground trembling under your feet, and you may yet see, crossing the riven flanks of the braes above the pines, the wraiths of the Children of the Mist, and hear the baying of Glenorchy's black hounds.

– Marion Campbell, *Argyll: The Enduring Heartland*

THE RIVER ORCHY is as short-lived as it is greedy, as it is beautiful. It springs into life fully formed from the womb of Loch Tulla, and succumbs fifteen eventful miles later into the north end of Loch Awe. No yard of its dark and wilful surge south-west through Glen Orchy and Strath Orchy is dull. In spate, as Marion Campbell observes, it carelessly plucks hapless alders from its banks and hurls them into oblivion, yet there are always miles of bank-lining alders, always new recruits to thicken the ranks. It whitens over rapids and sprawling waterfalls that present long-rather-than-high-jump challenges for salmon hell-bent on the spawning burns, and for white-faced downstream canoeists. It is unfailing friend to dipper, kingfisher, goosander, heron, otter and, increasingly, to ospreys – the fish-eating tribes.

And to me. I have known it, tramped its glen, camped on its banks, cooled off in its pools after long days on Glen Orchy's hills, for forty years. The commercial forestry industry has done some of its most thoughtless work in Glen Orchy, flinging dull green, one-dimensional forests over miles and miles of hillsides and brutally clear-felling in a landscape that deserves so much better, for Glen Orchy was a hearthstone of Scotland's historic native forest, and from here to the furthest north-east corner of Rannoch was one of the greatest flourishes of the so-called Great Wood of Caledon. Whenever and wherever nature can snatch half a chance to fight back in today's glen, it does so with alder on the low ground by the river and far up the mountain burns, with birch, hazel, aspen, willow, whitebeam, rowan, ash, oak and Scots pine.

Glen Orchy is also something of a watershed in the history of the Highland clans and right at the very heart of the unfolding events that made Scotland, the nation. A dear old friend, the late Marion Campbell of Kilberry, wrote a moving and highly personal interpretation of those times up to the very day when James VI turned his back on Scotland forever, in her timeless book *Argyll: The Enduring Heartland* (Turnstone Press, 1977):

This was MacGregor country. They trace their descent from King Gerig or Griogar, grandson of Alpin and nephew of that Kenneth who united Picts and Scots. Their arms display a fallen pine with a crown in its branches... From these mountains they preyed upon newer clans with parchment titles to Clan Alpin's hunting grounds. Out of this wilderness they were hunted, driven to change their name and find surety for good conduct, or else hang. Their women's faces were branded, their children herded into camps run by their enemies. These children could be flogged and branded for trying to escape; a second attempt (or one if they were over eighteen) was punishable by death.

This bestial business was King James's "parting gift to Scotland", in the words of the editor of the Privy Council Register. All chiefs with lands bordering the MacGregor country from Perthshire to Loch Lomond took part in the policy, but the experts were Argyll and his kinsman, Glenorchy. For many ghastly years the name of MacGregor was proscribed, until at last there remained twelve men who had neither bought safety nor changed their name. They were saved by a dispute between King and Earl over sharing the proceeds of fining those who dared help "the clan that is nameless by day".

The wonder is that any of them survived, but they did, much as their language survived the Act of 1616 for establishing schools "that the vulgar Inglische tongue be universallie plantit and the Irische language, quilk is ane of the chief and principall causis of the continewance of barbaritie... abolisheit and removit".

So much for the tongue in which Columba preached. The MacGregors were not the only victims of royal policy; the King in his wisdom had found a Hitlerish solution to several bothersome problems...

The River Orchy itself, its nurturing hills, a handful of the oldest trees... these are all that remain that also bore witness to all that. The river is the lifeblood of all Glen Orchy's eras these last ten thousand years since the Rannoch ice cap's benevolence gave it life. Its setting and its place in the hearts and minds of the people of the Highland west deserve rather better than it has been accorded by my own generation.

Glen Orchy is, of course, an eagle glen. Golden eagles often hunt ptarmigan and white hares over the plateau summit of my favourite mountain hereabouts, which is to say one of my favourite mountains anywhere. The views reach from Jura and Mull in the west to Rannoch in the east and Ben Nevis in the north, and back to Stob Binnein and Ben More, those Siamese twin mountains that straddle the divide between Balquhidder and Glen Dochart. Eagles have connected all this land forever. The golden eagles of Glen Orchy have never left. The sea eagles of Glen Orchy went the way of all their tribe, in much the same way that the MacGregors were eliminated, and by the same kind of people and for the same reason – they simply disapproved of their right to occupy the land. (How did you put it, Marion, "a Hitlerish solution"?) As yet I am uncertain of whether sea eagles ever nested in the glen, although I can't think why not, and I do know now that historically they nested on the islands of Loch Ba and Loch Tulla to the north-east, and of Loch Awe to the south-west, which means that for thousands of years Glen Orchy was at the very least a lifeline between sea eagle territories, and its river and open forest and hills were hunting terrain for nesting birds from all three lochs. And as the coast-to-coast eagle highway evolves through the early decades of the twenty-first century, the glen begins once more to resemble its historic past in this regard – that prospecting sea eagles from the east coast and the west are finding their way by accident or design to their historic homelands at both ends of the glen, and there they will find in Glen Orchy the familiar company of golden eagles and ospreys, and the wraiths of the Children of the Mist.

I never met the wraiths, despite many mist-bound days in the glen, but I had often met the golden eagles on the hill high above their nesting corrie. And once, after a long, hot September day in the hills I was slaking my thirst in the lower reaches of the Allt Ghamhnain and cooling off in one of its dark pools when three golden eagles appeared together gliding down the open flank of the corrie and almost wing-tip-to-wingtip-to-wingtip, two adults and their offspring of that year. Floating on my back in the water I raised a hand in greeting and gratitude, for their appearance was like a seal of approval or a blessing. "I do believe in God," said the American architect Frank Lloyd Wright, "but I spell it Nature." And amen to that, say I. And once I saw those eagles' kin from the neighbouring territory somewhere beyond the far side of the glen slashing a contour across the autumn yellow hillsides and banking one after the other into a side glen, which they sped through at a climbing angle until they burst from shadow into sunlight again, pulled back on nature's joystick (or however nature effects such things) and climbed vertically, shining copper and gold, until they dwindled to black motes of dust, and if you pushed me I would guess they were about five thousand feet up by that time.

The River Orchy is almost serene where it emerges from Glen Orchy and curves west into the wider, quieter plain of Strath Orchy, its havoc wreaked, its turbulent youth far behind it, likewise the roar of the "angry ghost in a long-severed Celtic head". Ahead is the steely Loch Awe that the river greets with a last left-handed swerve to the south. Loch Awe lies on the land like an uncoiled snake, its jaws agape and clamped round the landmark mountain massif of Ben Cruachan. Its tail is twenty miles away in the south-west, and no more than a few minutes of eagle flight from the sea at Loch Craignish; and beyond, a cluster of small Hebridean islands – Shuna, Luing, Lunga, Scarba – whose skies are dominated by the more robust glories of Jura and Islay. Loch Awe brings all that finally within the grasp of a young sea eagle that took its first faltering flight among the distant, gentle greenery of north Fife where it crossed a low ridge, saw the Tay estuary for the first time and recognised both a source of food and a highway west.

Loch Awe has been a mainland haunt of wandering young sea eagles from Mull from the very earliest days of the whole sea eagle adventure, and historically it was part of their heartlands, and now, it opens up the west for birds from the east. But westering birds that decline the loch's south-west diversion hold to Cruachan's southern flank above the open jaws of the water snake, thread the defile of the Pass of Brander and emerge into the bright new world of Loch Etive, and it too is an arm of the highway, for some of its headwaters gather in the hills above Loch Tulla. The final link in the chain is in sight, for there, sprawling across the western sky are the heaped mountains of Mull, and beyond the mighty Sound of Mull, the treasures of Sunart and Ardnamurchan.

☉☉☉

West of the Corran Ferry, south of Ardgour, and east of the Ardnamurchan Peninsula, lie the hallowed lands of Sunart. Hallowed, for there you will find astounding vestiges of the coastal oakwoods that once shielded much of Europe's western edge against the slings and arrows of Atlantic storms. The Sunart oakwoods are glorious, their survival is miraculous, and their future is guaranteed thanks to enlightened management by today's human natives, management that embraces conservation, careful timber extraction, a new generation of workers-in-wood, and tourism with a refreshingly light touch. All that has a name, a millennium project called the Sunart Oakwood Initiative. In a chapter about these woods in an earlier book, *The Great Wood* (Birlinn, 2011), I wrote:

Nature of course, had been working the woods for rather longer, and changes wrought by climate, fire, storm and flood, all manipulated a forest like this, discouraging some species, washing up new ones, readjusting the ground cover, the insects, the birds, the mammals. The oakwood I walked in on the first day of spring is a much-tampered-with piece of ground, for all its aura of timelessness, and the tampering goes on to this day, although ours is arguably the first of all this wood's eras to approach it in a mindset of healing.

The mindset of healing is infectious. I found myself involuntarily slowing my footsteps and commanding them to soften their sound and their imprint, evolving an exaggerated gait that suddenly reminded me of a Tom and Jerry cartoon in which one or the other is trying to sneak past the sleeping brute of a guard dog. So I told myself to relax and walk properly, but softly and lightly, naturally.

Loch Teacuis is an arm of the Atlantic that cuts deep into these woods, and opens into their main artery, Loch Sunart. I prowled the woods and the open shore near the loch's narrowest point where the sea fairly rumbles inland on a rising tide. A falling tide is generally better for otters, but "generally" does not take into account the possibilities offered by that other crucial element – luck.

Mostly, I have acquired my personal store of knowledge by repeatedly reworking the same set of circumstances until something close to intimacy begins to enter the equation and I begin to feel more like a part of the landscape than an occasional visitor. But from time to time, all is superseded just by the fluke of being in the right place at the right time. Such a day is part of my personal anthology of the Sunart oakwoods. I had a vague plan to walk through the woods to where Loch Teacuis melts into Loch Sunart, then return by the shore of Loch Teacuis to the place where I was staying. A good oakwood – a good wood of any kind – mixes different densities of trees with clearings, water both static and on the move, bogs, low-lying glens and dens and high vantage points, airiness and brightness with cloister gloom. If such a wood throws in occasional glimpses of mountains and the scent and sound of the ocean, it is fair to say that I will be that wood's happiest bunny.

One of the vantage points that day was a grassy and almost treeless mound that rose seventy or eighty feet above the woodland floor. It offered a sightline into a particularly dense copse of youngish trees fringing an older core where the trees were dark and thick of girth but not too tall. Ever since I saw a goshawk a few days before and only a few miles away, I took time to inspect every change in the terrain with binoculars before I plunged in. A formless fretwork of branches belonging to a hundred different trees takes time to inspect, but I had time. So I sat down and worked the glasses from canopy to floor and from east to west, examining everything that caught my eye, covering the most promising stretches more than once to make sure. Finally the glasses came to rest on something bright pink.

Such a shade is not an obvious one to stumble across in an oakwood on the first day of spring. It puzzled me at once, and I could make nothing of it. It also annoyed me, for I had already decided that anything pink and inanimate in an oakwood must be litter, yet these are the kind of woods where there is no litter at all. I stared at the thing with a "What the...?" frown furrowing my forehead. So it was pink, and small, no more than a few inches across, and roughly triangular but it was also partly obscured by branches, so not necessarily triangular at all. It was deep in the copse and high up among the branches of one of the bigger trees. Then it moved. It became blurred for a moment then resumed its place, as if a gust of wind had bowed the branch where it had snagged (I was still thinking "litter") then blown on by, so that the branch also resumed its position. But there had been no gust.

I picked out a tree on the edge of the copse in a direct line with the pink thing and set off downhill. When I reached the tree there was no pink in sight, so I headed into the copse along the same line, looking up. Then, as I neared the biggest trees, it reappeared straight ahead and about thirty feet up, and this time it was very clear what it was – a pink plastic wing-tag, and it was attached to the wing of a very large sea eagle indeed. At the precise moment I made my diagnosis, all I could see of the bird was most of one wing, most of the tail (no white, so a young bird, probably last year's chick from a Mull eyrie on an away day to the mainland), and one foot that had apparently been clumsily strung with fish hooks instead of fitted with regular-sized talons at birth. But then nothing about a sea eagle is regular-sized. Then a shudder went through the bird that burst open folded wings and pushed the head forward so that it now showed beyond an intervening oak limb, and I saw in profile the hook of an oversized bill and that hooded eye, which even in the deep shade of the copse seemed to be lit from within.

And of course that eye saw mine in the same instant, and I was unmasked at twenty yards. The eagle leaned back and was erect again, and for a moment the wing-tag was back where it had been a few seconds before. Then there was a new convulsion and the bird was suddenly airborne, except that there was almost as much wood as air in that unfathomable weave of twigs, branches and limbs. Unfathomable to human eyes, but fathomable enough even for the most juvenile of eagles, for it made a tight turn with its wings almost vertical for a moment so that it looked briefly like an oak tree itself, then the whole edifice righted itself in a single, impressively fluid glide and slid smoothly down to a yard above the woodland floor, and bore clear out of the trees below the level of every encumbering branch. The woodland confines magnified the bird and the huge gestures of its wings, and dignified the accomplishment of its short flight through the copse.

Out in the clear air, the eagle climbed steeply above the bank of trees that lay between the mound and the hidden shore, then I saw it veer away up the loch. For no reason that makes any sense to me now, several years after the event, I ran after it with what little speed was available to an earth-bound mortal over such terrain – out of the copse, across the clearing, up the slope of the bank of trees, across its wide crown, then steeply down the far side until I burst out of the oaks and onto the rocks and mud of the shore of Loch Teacuis where I stopped long enough to recover my breath and the shards of my composure, and to look up the loch for the eagle. There was, naturally enough, no eagle in sight, but only because I was looking in the wrong place.

I had scanned miles of empty high-tide shore in both directions and on both sides of Loch Teacuis. I looked for the eagle, for any eagle at all. Finding none, I looked for otters (fat chance at such a high tide), for sometimes sea eagles try and rob otters of fish they have caught and brought ashore. Finding none, I looked for crows displaying their "eagle-alert" behaviour, which in my experience in landscapes like this is indistinguishable from their "otter-alert" behaviour except that sometimes altitude is involved. Finding none, I fell back on my well-practised ritual for situations like this when something rare and extraordinary has happened and been superseded by nothing at all: I find a flat rock and sit down.

Having sat, I tried to come to terms with a niggling sense of unfinished business. Sometimes this happens. Sometimes, when it does, I respond to an internal command that I should sit still and wait, in expectation of more to come. Logic plays no part in the process. Instead, I start to scrutinise the circumstances I have just gatecrashed, in this case my slightly absurd pursuit of inexplicable pink. That scrutiny concludes – no, it does not conclude, it inclines towards a possibility – that the scene I encountered was but a moment in an unfolding sequence of events that had yet to run its course. There is no common ground and no pattern to the occasions when the phenomenon arises, and sometimes it is well founded and others not, in which case I feel like an idiot after a fruitlessly expectant hour of sitting and waiting and seeing nothing at all. But it all has to do with trusting that particular instinct which is a kind of shorthand in my writing life for all the years of watching and wandering and tracking nature's spoor and coming to conclusions about what I find and then trying to write them down. So I sat.

I sat because I heeded the command which I have come to think of as nature speaking to me directly, and when I can contrive nothing at all in response to nature's urging I tend to think that the failure is mine, that the quality of the waiting and the watching is lacking. I do not like to disappoint nature so I try harder, sit longer, wondering "what next?"

So once again, I was sitting and painstakingly scouring all the ways I think an eagle might have flown from its perch in the oakwoods, and I was finding nothing because I was looking in the wrong place. Across the loch and further inland from the narrows and away to my left, the dark wall of hills relented and leaned back and climbed less steeply, so that it had a more open, brighter aspect. A buttress of black rock guarded the near corner of that brighter land. Suddenly the airspace beyond the buttress was smothered in sunlight, crammed with sunbeams aslant, a moment so startling, so vivid, so golden, that I stared at its sorcery. But there was something else up there, a new movement among the shifting, translucent diagonals of sunlight. That yellow-gold air was violently astir with eagle wings, with the wings of no less than four sea eagles. An aerial ballet was in mid-performance, a thing of fluid grace and beauty executed by birds which until that moment I had never suspected of possessing such possibilities. Overlapping wings curved and filled like sails in a rising wind, and the four birds rose and fell, turned on vertically held wings, touched talons, and finally (in a gesture that reminded me at once of the Glen Orchy golden eagle family) formed a level, gliding line that drifted away west over the hills and far beyond that outpouring of sunlight. For perhaps ten minutes more the sun blazed down on an empty stage, then the beams faded and they left behind a still, grey hillside.

One last glimpse of the birds showed them high and dark and heading line astern across the Sound of Mull.

⊚⊚⊚

Ardnamurchan is out-on-a-limb Highlands, more like an island with a land bridge than an oceanic thrust of the mainland. It was the childhood home of a great friend, the nature writer and wildlife photographer of distinction, Polly Pullar. It is a place to which she still returns often for its particular savours of wildness, for its sustenance, and – in the manner of all true migrations – for its sense of destination and home. She has an awareness of and empathy for the needs of nature, from ecosystems to individual creatures, and a gift for healing nature's casualties of which I am deeply envious, for I have not a trace of it. Wherever I have paid tribute in these pages to the benevolence of "the grapevine", it is as likely to mean Polly as anyone else. She was the third of the four friends I went to for an account of their first sea eagle encounters. This is what she wrote:

Many wildlife moments are stored in the annals of my mind in a file marked "epic". My first sea eagle sighting is certainly in that special place. I was sitting on the shore at Achateny beach in Ardnamurchan, a wild and windswept bay overlooking Rum, Eigg and Muck. Fleeting rainbows danced over Rum's high peaks as a grey-splodged rain fell fizz-like on a sea of pewter. I had been watching a pair of otters on a barnacle-crusted rock, tugging at a butterfish, emitting their shrill, whistling contact calls to one another. I was fascinated by the way the water ran off their pelts, and the sound of sharp teeth through fish bones, accompanied by the melancholy cries of a curlew. The bladder-wrack swayed gently, engulfing the two lithe creatures and they vanished, but I sat a while letting the lovely sighting live on a little longer. I was quietly basking in my reveries when a movement on the shore caught my attention.

There was something large hunched over the remains of a dead sheep. At first I thought it was a golden eagle but it seemed so much bigger. Its almost lazy appearance had the air of a huge vulture. I could only see its back through my binoculars as it pulled vigorously on the carcase. A hooded crow appeared and dive-bombed it cheekily. I was struck by how tiny the hoodie seemed in comparison. The big bird kept feeding but was clearly irritated as it ducked its head and tried to avoid the unwanted intrusion. Then it suddenly roused its feathers, sending beads of moisture off in a spray, and effortlessly lifted off the sand. The sight of massive talons made me gasp. It was a sea eagle.

It skimmed across the sea, great wings outstretched in a stupendous eternal glide. Within minutes it was almost at the vertiginous sea cliffs skirting Rum, then lost to view through my rain-fogged vignette. It appeared so integral to this magical island seascape, so much at one with the setting that I almost *cried. I had that lump in the throat that perhaps only a nature watcher truly understands.*

It was then that I noticed the tide was about to engulf me. I wandered euphorically up the beach to the birch woods where I sat down again with my back to a wind-sculpted stump. The sea eagle was back where it belonged and I was ecstatic.

Chapter 9

MULL

Under the Yellow Mountain
Warm winds filled the hopeful wings
of limbering-up apprentice swallows
on the corner of south Mull nearest Africa
as September ebbed and dipped
a toe in autumn's imminent flow,
and going-nowhere eagles leaned
on the same benevolent winds
to tilt the Yellow Mountain.

Come April, come the same
Africa-and-back swallows, time-served
on big game plains, not cold rains
and sodden sheep; they perched again
on the same old south Mull fence
and saw the same old eagles
still tilting the Yellow Mountain.

NOWHERE IN ALL SCOTLAND embraces, emblazons, proclaims, promotes, exploits and generally brags about its eagles quite like the Isle of Mull. I like to board the island by the side door, by way of the Corran Ferry a few miles south of Fort William to Ardgour, then the rugged drive among the hills and oakwoods of Morven to the precipitous coastal village of Lochaline, which is surely the most consistently and foully mispronounced place name in the Highlands. (It is not the teeth-grinding, grimace-inducing Lock-a-*line* or even Lock-a-*leen* but rather the sigh-like Loch-*aah*-lin, with long-vowelled emphasis on the second syllable and the last syllable having virtually no vowel sound at all. There is no charge for this service.) From Lochaline then, the ferry trundles thoughtfully across the Sound of Mull to Fishnish, a little nowhere on Mull's east coast with a slipway, an unpretentious café to cater for the patient occupants of the steady dribble of queuing vehicles; that and a great many unwelcoming Forestry Commission trees – although sea eagles rather like such plantations for nesting. But Salen is only five miles up the road with good coffee, a good shop, and a post office where a staff sweatshirt announces: "Isle of Mull – Where Eagles Fly". Just along the road a sign proclaims: "Sea Eagle Bookings, 200 yards". I have been on the island for about ten minutes.

Sea eagles have done this. It was on Mull – in 1983 – that two birds from the Rum reintroduction project first nested, and – in 1985 – reared their first chick. A far-sighted conservation strategy established not just protection for the birds but also nationwide publicity. You cannot hide birds this size with a tendency towards very public behaviour, nor nests that can grow over a few seasons to more than twenty feet wide, so you may as well shout about it. A visitor hide with a big screen and a camera on the nest were installed, a convoy system was introduced to conduct visitors' cars from the road end (the nest site was mercifully in a plantation forest, which permitted a discreet approach through dense spruce to the hide), and – crucially – an island-wide campaign was launched that swung the islanders behind the conservation effort and recruited hundreds of pairs of eyes to help to counter the worst intentions of inevitable egg-and-chick thieves. All that was allied to a tourism initiative that openly welcomed eagle-watchers, and is now worth several million pounds a year to the island economy. Mull's eagle population has become something of a golden goose.

Before the sea eagles came, Mull was still one of the best places to watch golden eagles and a range of wildlife, from otters and dolphins to deer and basking sharks and a breath-taking list of bird species, but no one was shouting about it. Now, the boat trips are the visible tip of an iceberg that hoves to on tourism's sea every spring. A particular species of visitor decants from cars and campervans at selected spots around the island where they set up tripods for telescopes and cameras with lenses the size of baseball bats. They know where all the sea eagle nests are, and where the eagle boats go and sit out on the water with their engines cut, and try to lure the sea eagles down from their cliff edges or their tree perches. The landlubbers crowded around the tripods are based beneath the flightpaths the birds use fairly reliably between their favourite perches and the familiar boat with its familiar cargo of dead fish, the skipper's practised patter, and more people with cameras with baseball bat lenses but no tripods.

The composite message of all that is unambiguous: eagles are good for you. I am slightly uneasy about its evangelical edge, and when I go to Mull to watch eagles and much else besides, I'm on a different mission. I don't need photographs.

I try and find what I'm looking for far from roads and tripod clusters and boats bearing gifts of dead fish. But that's just me. I work with what George Mackay Brown called "the scrutiny of silence". One of the joys of Mull is that it is accommodating enough for all of us, and I begrudge no one a single moment of the pleasures to be found in collective birdwatching, nor do I begrudge the islanders a penny of the spoils. As a public relations exercise on behalf of nature and people, it has a lot to recommend it, and there is room to spare on many a quiet headland for the solace seekers, the scrutineers of silence.

Mull has had longer than anywhere else in Scotland to tune in to the behaviour patterns of a resident sea eagle population. But around the now improbably distant Tay estuary, where there is no tradition in living or even historical memory of the presence of sea eagles, the natives have not known what to make of the untutored juveniles in their midst, and now that some of these have grown to adulthood and the full glory of the bird is apparent for the first time, I sense a softening of attitudes, a quickening interest, and that these have begun to redress the imbalance of widespread suspicion and outright hostility. On Skye, there are boat trips, there is an eagle-watch visitor centre at Portree (where I first learned years ago now and with some astonishment that gannets were a regular item on the sea eagle menu!), but there is nothing like Mull's well-mobilised goodwill-to-all-eagles campaign. Perhaps Skye still thinks of itself as a golden eagle island, and there are certainly parts of the island where crofters still air grievances about sea eagle behaviour. In Torridon, where fifty-eight birds were released between 1993 and 1998, some of the crofters' outrage has ancient echoes of badmouthing wolves. It became the new scapegoat on the block for every real or imaginary felony against sheep, with the occasional, and apparently inevitable warning about threats to children. Stories of eagles taking babies from the fields while the corn was being cut are as old and far-flung and ubiquitous as stories of wolves being thwarted from their meal of babies by heroic hunters. It is the same thing exactly, with the same amount of truth to it, exactly. Who would have thought that the sheep, shorthand symbol for the Highland Clearances, would become a twenty-first century sacred cow used to deride the repatriation of an exterminated native bird whose presence in the landscape predates the coming of the first people, a bird ultimately exterminated by the same Victorian philosophy that also exterminated hundreds of Highland communities; that a little over a hundred years later, the people would find common cause with the sheep rather than the restored eagle?

◎◎◎

The Treshnish Isles, endlessly recomposing motif of the west coast of Mull, are confusing from Haunn, confusing because the island-watcher views their ocean-going convoy from dead astern so that their slabby profiles overlap and merge into an amorphous island porridge. A late afternoon of mid-April, the sun blazes off the ocean in a yellow-white dazzle. The south-west wind has been building for a day and a half, and all afternoon shore-bound waves have advanced to its raucous tune. They form up in a restless queue off Haunn's battered shore of geos, caves, stacks, holed and broken headlands. A hundred yards out, the waves are ten feet high, ominously green and muscular. They rise in concave scoops then curve forward to envelop their own internal architecture in long tunnels. The architecture of waves is rarely revealed. Their nature is impermanence, temporary and rolling evolution followed at once by extinction, but that evolution establishes shapes, patterns, structures, of which the rolling tunnel is but one. I think of Frank Gehry. He could stand here and see buildings in the surf. His first sketches often look to me like breaking waves. Nature has been the source of much great art forever, and Frank Gehry's architecture is great art.

Now, even as the waves pile in, I can feel the wind begin to falter towards high tide, then as it slowly abates, the surge of warmth from the sun. The biggest waves break fifty yards out, charge the shore in snow-avalanche surges. White horses – who ever came up with that image knew a thing or two about waves. The people who lived here (their roofless townships are everywhere along this shore, just above the raised beach) were cleared, like the eagles, and left for Canada, New Zealand, wherever, with the sound of the white horses in their heads and their hearts.

White Horses

Far from that vacant green island shore
where the white horses rode home alone
they broke virgin land
and gave thanks

for journey's end,
for freedom,
for old songs worn
next to their hearts;

and having given thanks
asked only this: sweet grass
for the white horses
that rode home alone.

The unfurling of huge waves, the unfurling of huge wings, the casual deployment of raw power that breaks rocks, that breaks bones: Mull has a quality, a limitless capacity for grand gestures, a landscape worthy of eagles. It is, I concede, a poet's rather than a naturalist's response to the scene before me, but then I never claimed to be a naturalist.

The naturalist may be intrigued and breathlessly excited as the sea eagle tries to find its feet on the Tay estuary, but the poet may feel a tinge of disappointment that not every Tay-fledged bird has made an instinctive beeline for Mull, for a land-and-sea-scape that *works* for eagles. Yet some do. The first east coast bird to breed did so on Mull. Torridon birds have bred on Mull, abandoning the mountains, lochs, pine-woods and angry crofters of the north-west Highlands and bypassing the possibilities of Skye's eagle stronghold in the process. Even Skye birds have bred on Mull. And Mull itself has become a hub that radiates eagles to every compass point. It is almost as if the sea eagles have recognised not just the island's qualities but also (given their willingness to watch the people at close quarters) the islanders' determination to do the best they can for "their" eagles, and in the process, to make the most of what they quickly realised is their good fortune to have been chosen by nature to participate in a daring new adventure to which their island is uncommonly well suited.

◎◎◎

Meanwhile, the road sign at Salen had done its job well. "Sea Eagle Bookings 200 Yards." Days later its message was still niggling in the back of my mind and sometimes in the front of it. There is no tougher nut to crack in the matter of wild-life-encounters-by-invitation than I am. As I have indicated, I am a bit of a purist in these things, "aloner than thou" as a reviewer of one of my books memorably put it. Long before I began to write about it all for a living, I was a self-taught disciple of what Tom Weir called "the joy of discovery". You learn from your mentors and peers of course, but you do your own fieldwork, you go and meet nature on its terms and reach your own conclusions. That way, the knowledge you acquire has authenticity and you can trust it utterly, and it is worth much, much more than anything you cobble together from Internet sites, guidebooks, and... yes, the gimme-your-money-and-I'll-show-you-eagles school of ecotourism. But these very eagles were a part of the story of the book I was writing, so surely the boats that were a fact of their lives were part of the story too. And apart from that, I just wondered what it would be like.

Besides, I have form. The BBC's Natural History Unit once paid me to go on a whale-watching boat in Glacier Bay, Alaska, and a few days later to stand – with a professional guide – twenty paces from an adult grizzly bear on Kodiak Island, and these had provided two of the most enduringly meaningful moments of my writing life. But on my native heath I tend to work the heath in native solitude.

I had seen the films and photographs that the eagle boats made possible by way of the low-tech expedient of motor-ing in to within a few hundred yards of one of those known coastal sea eagle perches, attracting a small blizzard of gulls with thrown chunks of cheap white bread and rolls, thus catching the distant eagle eye, then throwing a fish over-board. The theory is that then the fisher-eagle flies, hell-bent on the day's easiest meal, while the decibel level rises on the boat and gasps litter the air as thick as gull cries and an eight-or-nine-foot wingspan fills the frame of camera screens and viewfinders.

And perhaps, as it flies off with its prize, its yellow-gold eye will seek out yours in a passing moment of timeless inti-macy and an old, old connection will slip into place and change your life. Perhaps. It happened to me in Alaska when that humpback whale's four-inch-wide eyeball slid past the boat a few feet below me looking up, and I felt chosen. So, I would discover, did everyone else on the boat. And it hap-pened again on Kodiak when a grizzly sow with cubs broke off from salmon fishing to advance towards me and I asked the guide what she was doing.

"Just taking a look."

"At us?"

"No. At you. She knows what I look like."

So now there was a roadside sign at Salen and I began to wonder about it, remembering Alaska, wondering what it might be like. I wondered what happens to the people, and to me, and to the eagle when it volunteers to mix it up with the species that wiped its ancestors from the face of the island, from the Atlantic seaboard, from the face of the land. So I bought a ticket.

The skipper was a jovial cove with apparently inexhaust-ible supplies of jokes, banter, bread and fish, and with very good birdwatching eyes. The male eagle had been "in great form all week, really putting on a show, no reason to think today will be different, and let's hope you'll have a day to remember all your life".

Naturally, he added a professional disclaimer:

"But remember these are wild birds and I don't have a rope I can pull to call them in." Now why would he say a thing like that?

He said he thought the male recognises the boat, and even the sound of its engine, knows the routine.

The morning had been sprinkled with magic dust. The sea was the blue of alpine gentians, all but flat, and littered with winking diamonds of sunlight. It grew warm and windless. There was high snow on the summits and ridges of Ben More and the other high hills, just enough to confer an air of alpine distinction on them. The west coast of Mull in this mood may be the most beautiful place on earth. Suddenly what we were about to do seemed ridiculous in the face of so much beauty. For some reason I thought (for the first time in about forty years) of Lillian Beckwith and what she might have made of ecotourism, the preposterous idea of a boat full of tourists being so easily persuaded to part with their money so that a man in a boat would take them to see eagles. Maybe see eagles.

So our little group sailed out with high good humour and higher expectations. Eventually we nosed gently into the head of a sea loch and cut the engine. Silence roared in. Two red-throated divers sat unperturbed on the water not thirty yards away. We already had a retinue of gulls thanks to some thrown bread. The skipper announced that the male sea eagle was exactly where he had expected it to be, and identified a smudge on a distant conifer. The longer lenses on the boat had a speculative look, but even in good binoculars the bird looked like a bit of a dead branch.

We waited. We watched. The bird's slightest movement was announced out loud by any of us, by all of us, a lifted foot to clean its beak was met with an "ooh", a half-open wing raised a louder murmur, a shuffle along a branch almost a shout. It was getting tense.

A fish went overboard. Two big black-backs disputed ownership and fifty lesser gulls complained. The divers edged closer towards the fringes of the melee, looking for scraps. The eagle stayed put.

An hour passed, a singularly agreeable hour, what with the weather, the setting, the birds. Up in the big conifer, stupor had broken out. We ate sandwiches, drank coffee, waited. The skipper was philosophical and apologetic by turns, threw more fish, created more gull mayhem, all of which the eagle in the tree ignored. One of the divers took off with a runway of about a hundred yards, flew in a low arc that brought it round on a collision course with the boat, but it climbed at the last moment and hurdled the boat not ten feet over my head. I heard my own gasp. What was that all about?

I turned back to the ship's company to share my delight, but they were all looking the other way.

"It's flying!"

It was indeed. The boat galvanised. The biggest lenses were trained, primed to follow every yard of its approach to the boat, another fish went overboard, another undignified scrap among the gulls, and in the mind's eye of all of us at that moment was the prospect of a sea eagle gatecrashing the throng to snatch the fish from under their screaming noses.

The sea eagle was coming our way, eventually. But first, and to the collective dismay of the ship's company, it began to circle, and as it circled it climbed, and as it climbed its already distant profile grew smaller and smaller. Finally, once it had put five or six hundred feet of airspace between itself and the sea, it levelled out and headed south out across the loch and it did fly directly over the boat and over the not-quite-long-enough lenses.

Oh, but there was something god-like in its passage! I see it in my mind's eye to this day, a single indelible image, the bird brilliantly lit against a royal blue sky, its white tail "shining like the silver moon" in John Colquhoun's unsurpassable phrase, and those wings at their mightiest outreach – God, what wings! The bird was cruising, the wings moving to a slow and shallow pulse and held intermittently in brief glides, and it was headed for the hills that cluster like acolytes around Ben More, the Great One. And if you had the sensibilities towards the natural world around you of, say, the tomb builders of South Ronaldsay, and you were accustomed to seeing such a sight on a daily basis, knowing that the same bird would come down to your level, lift a hare from your rough field or a fish from the bay where you fished yourself... why would you *not* make a god out of such a creature? Why would you not invite such a creature to participate in your sacred rituals following the death of your kin and your friends?

That's what occurred to me, standing in a small boat on a flat-calm sea loch one mid-April day. It was the last thing I expected to feel, but it means rather more to me than if the bird had come down and taken a thrown fish in front of the cameras.

And that was it. Except that half an hour later the bird came back, higher over the boat if anything and followed by a second eagle which may well have been escorting it from the premises of the territory of a second pair where it had just trespassed. That, at least, was my best guess.

Two other things strike me now about that day. One is obvious and salutary: the recognition that you cannot orchestrate nature, that wildness by definition means unpredictability, driven by impulses that are different from those of our own species that turned its back on wildness. The other is that no one complained. No one felt short-changed, no one was dissatisfied, no one objected to being outwitted by an eagle, to being a bit-part player in an eagle game. "Game" is the wrong word of course, for it implies intent on the eagle's part, whereas its behaviour was almost certainly down to the fact that it wasn't hungry enough when we showed up. But watching it cross the loch and refusing the lure of our very obvious boat with its obvious gifts, a game is what it felt like. Yet the laughter and chat on the way back were evidence enough that we had all won, we had all taken something positive out of the game.

An eagle that conjured god-like presence, a red-throated diver that flew over my head, the west coast of Mull in such a mood... the skipper was right, a day I would remember for the rest of my life. I thanked him for the roadside sign at Salen.

☉☉☉

There were strange sequels to my brush with ecotourism. The very next day, walking the shore at Haunn where I have walked dozens of times over perhaps thirty years, I was stopped in my tracks by a rock I had never noticed before. It may just have been that the light was particularly good, or that the state of the tide revealed the whole rock in all its aquiline glory, but I found it hard to believe that I had never given it a second glance before. It's not as if it is inconspicuous. It is perhaps forty feet high, perhaps sixty feet across at its widest point, and in my eagle-crammed mindset, it resembled nothing so much as a sea eagle standing erect with its wings half open. The rock was black above a very obvious high tide line where its "feet" might have stood, and below that, the rocks that were accustomed to being underwater for half the day were a pale grey. The contrast heightened the impression of a monumental eagle set on a plinth.

It could be, of course, that I saw it in such a light because the boat trip was still so fresh a memory and because this particular visit to Mull was part of the process of nailing my theory about the eagle highway into place. But even now as I write, a year after the event, the photographs I took and the crude drawing I made are completely convincing: a massive representation of an eagle, sculpted by nature on the shore of the island that is my idea of the highway's destination. How could I possibly have missed it before?

Then, a few hours later, there was an encounter with a young shepherd who was helping out with the lambing for the local farmer. He is an affable young man I had met before, a fifth generation Mull shepherd with a keen eye and a good feeling for the land of his ancestors. The conversation came round to the wildlife he lived with on a daily basis. He said he liked birds particularly, but not the sea eagle. There were too many of them now, the change was too fast, and they were killing all the wildlife. He said there were no hares now, and that sea eagles kill the swans' cygnets and the protected seals out on Colonsay. His accounts were based on circumstantial evidence and anecdote, but he was clear that such instances were widespread and frequent, and that the net effect was profoundly negative. His was the first disapproving voice that I had heard on the island. He would not hear a word against golden eagles, but the sea eagle project had gone too far.

The next evening in the pub I got talking to a couple of islanders. They had moved up from Lincolnshire three years before and seemed to me to have taken to island life with an infectious zeal you don't always find among what the natives call "white settlers". Once again, the subject of eagles came up. They lived in a tiny community in the north of the island and went into raptures about how they could watch two sea eagles from their bed! They also described to me at length and with the same rapturous expressions and gestures the sight of three sea eagles, a golden eagle and buzzard riding the same thermal at the same time.

A loud-voiced couple with London accents had made our conversation difficult at times, and the odd phrase that drifted across provoked knowing looks from the eagle-worshippers. The London gent had been briefly eavesdropping, it seemed, for he suddenly leaned across and said:

"I saw an eagle today!"

The Lincolnshire-lady-turned-islander smiled appreciatively and responded:

"Golden or white-tailed?"

"Pass," said the Londoner. "It was sat on a telegraph pole."

Then he dived headlong into his sticky toffee pudding with butterscotch sauce.

The relationship between people of all kinds and the eagles of Mull of both kinds is a complex one. What has made the difference in the time that I have known the island is the impact of sea eagles. My first visits to Mull more or less coincided with the first sea eagles' arrival on Rum. It would be years before I started to see them. But I registered quickly and with delight that Mull was the best place I knew to see golden eagles, and I quickly learned a handful of good places to walk and be still and expect to see them in the course of a day. I never had a conversation with an islander about golden eagles unless I brought up the subject myself, which I rarely did. The few responses I can remember were either that the islanders in question had never seen a golden eagle or else there was a quiet, almost instinctive reverence for the bird expressed in words that suggested it inhabited, as one old man told me, "that other Mull". But thanks to the sea eagles and their habits of intruding on the everyday lives of the natives, and thanks to the tourist promotion, and thanks to the islanders' widespread embrace of every aspect of an "eagle economy" (I have heard the expression several times), now everyone I talk to seems to have an opinion or a story about the place of eagles in island life. Mull, as never before, is an eagle destination, and in a Scotland-wide or for that matter Britain-wide context, an eagle destination unlike any other. It is as true for the eagles as it is for the people.

◎◎◎

An old September, a dozen years ago now, a self-catering cottage in the early evening far out on the Treshnish headland of Mull's north-west corner, the door wide open while a meal was being prepared, a dram poured, the day mulled over.

The sounds of the island drifted in through the open door, the birds that constantly came around the house or drifted across the hill at the back and the headland at the front, sheep on the hill, cattle in the rough fields, wind, and a strange and barely audible high-pitched and feeble yelp like a terrier pup that had just pricked its nose on its first thistle. It was a sound at the edge of my hearing, but it stopped the early evening in its tracks. I switched off everything and went to the door, dram in hand. The terrier voice sounded again but there was nothing to see, not yet. I waited. I know that voice. Then, one by one, three golden eagles appeared round the hillside, no more than a hundred feet up, the first one noticeably more compact than the others – the adult male; the middle one particularly large and dark – the adult female; the third one lagging behind a bit and the one that was doing the talking while showing some fallibility navigating the thermals and cross-winds around the corner of the hill – that was the year's youngster, newly fledged and learning the ropes of flying from the past masters of the art, the birds that can land flying backwards, the birds that can play in choreographed overlapping vertical circles flying backwards at the top of each climb, the birds that can reach upwards of a hundred miles an hour just by half-folding their wings and angling down into a shallow glide, the birds that can hunt a foot above the ground at a hair above stalling speed, the birds that can snatch a tormenting crow from the air with a fast spin and an upwards strike from an upside-down position.

All these I have seen for myself. All these, plus the bird that was being mobbed by two ravens up behind the Glen Dochart cottage, and which struck one of them in the air so violently that it fell to the ground and dived down after the second bird and caught it in the air, then landed to pick up the first bird, and rose – much more slowly – for the three-mile journey back up the hill and over the watershed to the eyrie with a raven in each talon.

Such was the pool of expertise at the disposal of the young golden eagle with the terrier yap as it beat through the turbulent air round a Mull hillside, following its parents past the end of a hill track and over the roofs of three small cottages. They crossed the rough fields beyond, and headed out towards the blunt thrust of a headland to the north, and the yellowing sunlight of a September evening burnished their golden tints and made something immortal out of them as they passed, something symbolic that stood for all the eagles of all time. Ah, you had to be there.

⊙⊙⊙

An old April, early in my Mull years (spring and autumn soon became my preferred island-going seasons), wandering through the blunt innards of the island's north-west, where a gap in the hills revealed a far glimpse of Staffa. I have always loved to sit where I can see Staffa.

These are strange looking hills to mainland eyes, volcanically rounded and rising steeply to small flat plateau summits. The profiles of the hills skip down in tiers, each one flat and right-angled. The faces of the hills are banded by low cliffs, some only a few feet high, the biggest no more than fifty feet. They look like ragged barrels, and in their midst I stumbled on a hidden arena, like half a coliseum where the cliffs are carved across a curving headwall in four distinct leaps, the cliffs separated by level terraces hundreds of yards long and up to fifty yards wide, nature as architect. I found the place at all because I was lured there by eagles.

I was alone and dawdling over an empty single-track hill road. I had seen no one all morning, so I was driving at birdwatching speed, and looking for somewhere to leave the car that wasn't a layby and didn't involve crossing a ditch. It was a four-wheel drive car but not the kind that can cross ditches. The day was grey, the hill-grasses were that bleached shade of no colour at all that they acquire by the fag-end of a long wet and windy winter. The forecast suggested brightness later. I was equipped for a long day out based on the presumption that the forecast was wrong. I found a piece of dryish ground where the car could be abandoned, got out, looked around, scoured the hills, the sky and the middle distance with binoculars, and it was there – in the middle distance as usual – that I latched on to a golden eagle. That makes it sound easy. The searching had involved half a shivering hour of stillness, which is not as easy as it sounds, but the difficulties can be mitigated by choosing your time and place and trying not to look out of place.

The eagle was juvenile (it had white underwing patches) and probably male (my guess, based on its comparatively compact size). It may sound a preposterous idea to follow an eagle into the hills, especially hills you don't know, and in expectation of finding other eagles, a kill, even an eyrie – although there are very specific and stringent laws about how close you can approach one of those, and about disturbing nesting birds. (Besides, as I have already indicated, I have a pretty good idea what happens at a nest, but I am constantly astounded by what happens out on the birds' territory, and that is what sustains me, intrigues me, and sustains my writer's instincts.) But many years ago, as a hopelessly romantically motivated teenager, I read a book by Lea McNally about his time in Torridon, and about following an eagle into the hills. So I knew it was possible. It was just a matter of developing a technique I could rely on. From the point at which I read the book until the point at which I had developed such a technique, about thirty years had passed. If you want to understand golden eagles, you have to be in it for the long haul.

So I marked the bird's line of flight as carefully as I could until it disappeared over the lowest edge of one of those rounded hills, donned boots, jacket, rucksack with a day's rations, enough clothing to cope with a return to midwinter conditions and a small hipflask for ceremonial toasts to eagles or whatever else seemed appropriate. I slung the glasses round my neck, and picked up my stick. The stick is my eagle-watcher's secret weapon, made from cherry wood with a small deer-antler 'V'-shaped grip, the perfect low-tech combination of third human leg and a monopod for resting binoculars and occasionally a camera. And for leaning on for what my father would have called "a wee rest".

An hour later, I had seen nothing at all, but I had climbed an unpromising looking rise in the ground and found myself in the "wings" of the coliseum, so I sat and watched the landscape instead, to see what might rub off. The level floor of the arena was the size of a large field and was wall-to-wall bog. Beyond, the land seemed to spring backwards and upwards in four leaps to the skyline, the terraces and the leaps cut by waterfalls. Yet it was not a rocky place. The same winter-weary grass covered almost everything, blotched with dark brown withered heather and defeated bracken. No vestige of spring growth yet, not here, but after a while of sitting and looking and listening there were suddenly golden plovers calling and meadow pipits falling cock-tailed and singing down ramps of cold wet air. A knot of about a dozen red deer hinds watched me from the skyline.

Stillness in such a place rarely goes unrewarded if your senses and your mind are as busy as your body is still, and if your stillness is complemented by clothes that blend with your surroundings. Wildness is often what you make of it, but it is rarely orange.

The day drifted into afternoon, I sat on, watched my way through a slow, late lunch, and realised that it was snowing lightly. It is surprising (at least it surprises me!) how often a weather event stimulates some aspect of nature that has been still or hidden. Five minutes into the snowfall, a rock two hundred yards away and fifty feet higher up the hill suddenly changed shape. It changed because (I now realised) the golden eagle that had been sitting there for who knows how long, betrayed itself by vigorously ruffling its feathers to free them of snow then reassembling itself into an eagle shape. Then it flew. It executed a single low circuit of the coliseum, presumably for momentum and lift, then launched itself into a fast, steep climbing corkscrew such as I had never seen before (and have never seen since), but which was worth the admission price on its own. These days, I incline towards the idea that there is a high degree of individuality among golden eagles, and that in terms of the huge repertoire of powers of flight at their disposal, individual birds show a particular preference for individual techniques. My erstwhile friend and eagle authority Mike Tomkies wrote it memorably in his best-known book, *A Last Wild Place* (Jonathan Cape, 1996):

She came towards me from the east like some extra-terrestrial being, dark, massive and silent as a ghost, with an aura of the primeval about her. With her great wings angled back like thick fangs, she moved against the south westerly gale without the need of a single wingbeat but with consummate ease, as if owning a secret of aerial mastery no other bird possessed.

A secret of aerial mastery no other bird possessed was perhaps what I had just witnessed. Mike Holliday's reversing pair in Glen Ample was perhaps another manifestation. The more you take the time and trouble to wander out into landscapes like these, the more you will learn and the more you will marvel at what can unfold when a golden eagle puts its mind to it. On a clear day, the event would have been magical enough, but this thin gauze of snow conferred something arty and surreal on the moment, for a watery sun was now conspiring with the snow so that the light was filtered and sprayed all across the corrie, something like the way street lights splinter in fog. The eerie effect conferred a flimsy, ephemeral air on the landscape so that the eagle was the only thing of substance in my world, and all the more miraculous for that.

The purpose of the climb (if not the technique) was suddenly revealed – two more eagles, much higher, perhaps as much as a thousand feet, one substantially larger than the other, the mood apparently combative. I watched with the naked eye rather than the binoculars to keep all three in sight at once, and binoculars are not at their most useful in falling snow. The spectacle drew me in, my mood transformed from relaxed to heart-in-mouth. The cork-screwing eagle climbed far out into the mouth of the corrie, then spun from the apex of the climb into the wind and attacked. ("Great wings angled back like fangs" – perfect!)

The other two eagles had tumbled down the sky and were now much lower than the new attacker, whose fast glide turned into a free-fall aimed pointedly at the smaller bird. The falling bird pulled out as the target bird dodged, but it only pulled out into a stupendous upward swoop of about five hundred feet, flipped over and fell again. But now there was no target, for it had fled far across the island at formidable speed, and suddenly there were only two eagles in the sky, and these formed up wingtip-to-wingtip and disappeared round the end of the coliseum in gentle, level flight, leaving me ever so slightly breathless.

I reconstructed the incident as follows. Somewhere nearby there was an eyrie. The eagle on the rock was the resident male. Of the two higher birds, the larger one was his mate, roused to repel the attentions of an intruding male, the young bird I had first seen from the car, then followed into the hills. The bird on the rock had responded to his mate's alarm, although how that was transmitted is the business of eagles and quite beyond me.

I had watched the whole thing through the snow veil. I shivered and stood, and set out to explore the coliseum's ridges and tiers. The sun re-emerged, the snow faded and fizzled out, and a new and glittering clarity drenched my portion of Mull. I climbed first to a watershed where bare patches of dark and cindery volcanic rock looked like transplanted scraps of Iceland. A stone I picked up felt much too light for its bulk, porous and burned out by its volcanic origins, and dumped there to be stumbled over by deer and sheep, shadowed by eagles and fingered by my passing hand. The wind hammered up the headwall, as loud as it was cold, with ice in its breath. Its lulls were filled with new snow flurries. I was now level with the highest tier of the headwall, and rather than plod on through the wind to the succession of small skyline summits, I cut along the level terrace and found it magically shielded from the wind, and suddenly the walking was idyllic. For I turned and saw the islands and the fast squalls that tore among them, snatching them from sight one at a time then reinstating them as they raced on. And then, where I had paused to examine a tumbledown of boulders that lay heaped on a gentle slope, I lifted my eyes again and saw Staffa snared in the arc of a rainbow.

◎◎◎

There would come a grey day in the far north-west of the island when I walked a clifftop in a wet north-westerly gale, a day when one more grey rock among thousands of rocks caught the nature writer's eye because there was something not right about it, so it demanded a second glance. The wind was thrusting moist gusts over the clifftop from below: you could almost see the air burst apart into upwardly mobile columns, then fold over as they cleared the cliff edge, then go dancing away across the wide tilting plateau of the land, scattering mayhem in madcap polkas. I felt almost stupidly invigorated in the face of so much supercharged weather, half-inclined to dance myself, which (as those who know me well will confirm) is a singularly improbable circumstance. But then, advancing to where I could see low-tide rocks far below, there was that one awkward rock, awkward in the way that it related to the thousands of others.

So I raised my binoculars, flicking the rain-guard from the eyepieces in the same movement, and of course the lenses were pockmarked with raindrops at once. Yet through that gauze of glass and water, I could still make out the sideways and upwards twist of the head, the suddenly revealed bill-hook beak, and the upward-canted eye that took in my newly arrived shape silhouetted on the skyline. Then I saw the head twist the other way, twist so far round that although I was looking at the right side of the bird, I was suddenly being scrutinised by the *left* eye. Then the whole body of the "awkward rock" heaved in the same twisting motion as the head had done, so that the whole bird now faced into the wind and in that same moment became hugely airborne, an effort of such a brute mass of wings to convey the bird into the air. I thought again how a golden eagle would have done it, how in the same circumstances it would have stepped up onto the air and found lift with perhaps a single upstroke then half a downstroke that locked into a level glide with wings horizontal and still, how it would pass by and vanish. And I knew in the same moment that the sea eagle, though temporarily hidden by the cliff, would be coming my way.

Moments later, it appeared above the cliff edge and fifty yards to my left, facing towards me but drifting sideways over the land led by its starboard wingtip, although "tip" implies something altogether too tightly controlled for that colossal over-provision of over-sized primary feathers that cluster so thickly around the business end of sea eagle wings. I have held *one* such feather in my hand (at an organised RSPB talk: it was long-discarded by its original owner and was being used by the speaker as a prop). It was the length of my hand and forearm. I swatted the air with it, felt the air rock against my other hand.

Then the eagle stopped its sideways drift and came straight towards me, and was directly overhead and about fifty feet up when it held almost dead still against the wind, then it eased forward again and slowly began to circle the clifftop, watching, assessing, making pointed eye-contact. The sky, like the land, like the ocean, like the day, was unrelieved grey, and against that the bird looked black. It was a young bird so there was no white tail to relieve the bulk of the immense silhouette. I have been very fortunate to see many eagles in many circumstances and in several countries, and never found them less than admirable, never less than thrilling, and whenever I tried to write them down I was conscious of a particular desire to do them justice. This sea eagle, which any decent field guide would probably categorise as "immature", was so confident in its edge-of-the-land world that it navigated the turmoil of that ocean-going wind among the cliffs, effecting the 200-foot climb to emerge at the cliff edge within two yards of the outermost thrust of rock, so intimate with the wind that it could pilot itself sideways under perfect control then gently propel itself forward, using the wind – instructing the wind it seemed – to position itself on the pinpoint of air directly over my head, and I refuse to believe that the manoeuvre had any purpose at all other than to create an unforgettable impression. No one standing where I was standing that grey and turbulent morning could possibly fail to be convinced of the rightness of that bird in that landscape. There was no sound. There hardly ever is. There were no tricks, no flying stunts, there was just the unambiguous possession of the air, the land, the edge of the ocean, and a lingering impression in the back of my mind of being in the presence of something as eternal as the rock itself, as the ocean itself, as the very air it could command to do its bidding. Was it a moment (or many moments) like this, I wonder, that persuaded the earliest Orcadians to engage with these extraordinary neighbours, an engagement memorialised for us to marvel at five thousand years later on South Ronaldsay?

The eagle beat away out over the sea at an angle to the wind, and I watched it until it was too small to follow with my eyes (the binoculars were now less than helpful). If it held that course, it would be on Skye in minutes.

Skye!

Of course!

When the first of the reintroduced sea eagles began to fan out from Rum, both Mull and Skye evolved more or less simultaneously into breeding strongholds. Forty years later that status is more secure than ever. My eagle highway envisaged a destination on Mull, a journey's end for east coast birds. But what then? More will undoubtedly breed on Mull, but not all of them. So if not Mull, where next? En route to Mull from the east they have probably encountered Mull birds already, and as far east as Loch Tulla, Glen Orchy, Loch Awe, Loch Etive, all of them on journeys of their own across the country. Some of those east coast birds must sooner or later adopt the practices of birds from this long-established community of the west. So that also means wheeling away from Mull, perhaps with today's young island birds, to travel far up and down the west coast – to Kintyre, to Islay, Ardnamurchan, Skye, Torridon.

"Or Mingulay," I mutter to myself aloud, remembering that first rock that looked not quite right among so many others and became the first of all my sea eagles. But I have now crossed the country from east to west and back again many times, finding eagles or stories of eagles all along the way, and it suddenly occurred to me on that grey headland of north Mull that there is no end to the eagle highway. I could spend another year, another decade, wandering the western seaboard, and from time to time my path would cross that of wandering eagles and established eagles, and I could learn from that, slowly and over time.

But I have a history with Skye, which is longer than the entire sea eagle project, a history well populated with eagles for that matter, eagles of both tribes. I could let Skye symbolise for me all the eagles furth of Mull along the entire length of Scotland's sunset-facing, island-strewn, Norway-like coast, one last flight of fancy along the highway. I stared out along the course of the gale-deflecting bird, which had so recently paused in the sky just above my head, circled the clifftop, and then (the notion suddenly entered my head and I accepted it gratefully) pointed the way forward. I would go back to Skye.

Chapter 10

SKYE AND THE BROTHERHOOD OF EAGLES

THERE IS A PAINTING OF SKYE on the wall of the room where I work, and in the course of writing this book I must have lifted my eyes to it a thousand times. It shows Sgurr nan Gillean and Marsco from Sligachan and there is no more handsome prospect anywhere in the land. The painting is the work of Skye-based artist Duncan Currie, who also happens to be a friend. I got to know him through his father Andrew Currie who was the old Nature Conservancy Council's man on Skye for many years, and whose wildlife column in the *West Highland Free Press* was an institution within an institution. He was a man whose company, wisdom, knowledge, humour, friendship and whisky I enjoyed many times. Andy died in 2012 and I was asked by the family to speak at his funeral. On the day of the funeral I had a few hours to myself, drove up to Sligachan and walked a mile up the glen with those two mountains and the memories of Andy for company.

The next time I was in Duncan's studio I saw the painting and bought it, partly because it is a truly beautiful painting, and partly because it frames my memories of Andy in a singularly appropriate context. And the day that I bought the painting, Duncan told me his story about his first sea eagle, and I asked him if he would write it down for me. And quite unprompted, his written account moved seamlessly from his first sea eagle to his favourite golden eagle moment. This is what he wrote:

As you know, due to my dad's career in conservation, I had long been aware of the old NCC's plans to reintroduce the sea eagle to Scotland. So as stories of the bird successfully breeding became more common, I strongly desired to see one in the wild. However, years went by and somehow a sighting eluded me, despite much time spent fishing in the hills of Skye and the broader area of the north-west Highlands and Islands.

So it was with great surprise a couple of seasons ago when watching an approaching storm on a south-westerly gale from the window of my studio that I spotted what looked like a black bin liner wheeling wildly in the strengthening wind. It was coming my way and as there was something strange about its movements, I went outside to have a closer look. Imagine my joy when the "bin liner" transformed itself into a magnificent young sea eagle as it was buffeted over my head about sixty feet up, its feathers ruffled black "against the grain" by what was now a full gale. It was rapidly lost to view swept in the direction of Irishman's Point [a headland in the north-west corner of Broadford Bay], but I think it had probably been blown off course trying to get to Pabay [a small island to the north of Broadford Bay] as local fishermen had told me of a bird roosting out there at times over the past months. How ironic that after all these years, I walked no further than the door of my studio for my first view of a sea eagle. In fact I have seen them regularly since then but this remains my closest and most memorable sighting, so far anyway.

My most thrilling view of an eagle however – a golden eagle this time – was while fly-fishing for sea trout on a beautiful loch in the moors above Stornoway on the island of Lewis. I was fishing a wide shallow bay with quite a deeply cut stream running into it. I had worked my way a fair distance into the bay, keeping as close to the river channel as I could. Glancing back to watch my backcast, a movement caught my eye and I let my line drop to watch a low-flying eagle emerge from between the stream banks.

It was a stunning sight to see as the bird passed barely a cast's length away to my right, and almost at sea level. Completely unconcerned by my presence, it flew with almost casually slow wingbeats and it swivelled its head to watch me as it went. In that steady gaze, I felt a strange sense of communication as our eyes met. The bird, with no sign of fear, simply took in what I was doing, saw no threat, and passed on its way. My odd impression is one of professional courtesies exchanged between two "hunters" doing their respective things. If we could have spoken it might have been along the lines of, "yes, I'm having a quiet day too".

Over the years I have been privileged to have close-up views of many of our wild animals and birds while fishing – everything from otters to peregrines. I think fly fishers move slowly and quietly, blending into the landscape so they appear a natural part of it, therefore wildlife comes closer. Seeing that sea eagle filled a big gap in my "to see" list. Now, if I can just be in the right place at the right time to see that basking shark I've always wanted to spot...

⊚⊚⊚

In an endeavour such as this, and usually when you do not
see it coming, there dawns a morning supreme, a pivotal day
of revelation. Such days are as rare as diamond dust. This one
dawned from a web of days when the greyest shade of gloom
was as unrelieved at noon as it was at dawn and dusk, all of it
held fast in jaws of frozen snow and bare ice. The revelation
was all the brighter because I had Leslie Brown on my mind,
because I had been reading his book *Eagles* (Michael Joseph,
1955), and was struck by the fact that in a vibrant and wide-
ranging study of eagles in Britain and Africa its one note of
pessimism concerned the prospect of reintroducing sea eagles
into Scotland:

*A question that must be asked is, "Shall we ever have Ernes
breeding in Scotland again?" I should very much like to think
so, but I doubt it. The last Scottish Ernes were killed out more
than thirty years ago, and in spite of a changed attitude towards
large birds of prey, there has been no reappearance. Ernes are
seen now and again, but they are chiefly immature birds on passage from Norway, or Iceland, or Greenland, and when they
become adult and need to breed they go to those places to do
it. I believe that eagles return, generally speaking, to somewhere near their birthplace to nest, and as long as there were
some Scottish breeding sites of the Erne there was a chance that
they would survive. But now that there are none, and we must
depend on the reduced Erne populations of Northern Europe
to recolonize Scotland, I fear it is hopeless. They may pass
through, but they will not stay, and unless we can in some
way overcome the innate breeding mechanisms of eagles I do not*

believe that the Erne will ever be restored to Scotland. Golden Eagles have in recent years reoccupied some old sites, and likewise Buzzards have also spread, but they were British Golden Eagles and British Buzzards. One could, perhaps, take Erne's eggs in Iceland, fly them to Britain, and put them under a Golden Eagle, but it would be very difficult, for an eagle starts to incubate its first egg almost at once, unlike such birds as pheasants or ducks which lay a large clutch that you can carry away and put under a hen. Anyone seeking thus to restore the Erne would need to take an incubator with him to the Erne's nest, incubate the egg in the 'plane, and continue to do so until it was safe in the Golden Eagle's nest. And there is always the chance that the Golden Eagle might desert after all that trouble. Nowadays it might well cost hundreds of pounds to rear a single Erne in a Golden Eagle's nest, and how do we know that when the eaglet was grown it would not feel impelled to fly back to Iceland or wherever the egg had come from? It is a melancholy prospect altogether, for there are many places in Scotland where the Erne could do no great harm and would be a glory to watch.

Not sixty years after he wrote those words, how the eagle landscape is transformed!

So, it had been a long, long winter. But if you have kept up your vigil through the weary days, peering through the greyness in search of eagles that reveal only that they have gone somewhere else (or they are there but you don't see them because − like wolves − they sometimes choose not to be seen); if you have been willing to go back again and again in pursuit of the endeavour, then the diamond dust day is your reward. Suddenly there is sunlight, then rainbows, and beyond these, your pot of gold.

Twenty years after Leslie Brown wrote his pessimistic prognosis for sea eagle reintroduction, Rum and John Love proved him wrong. But he was right in one regard: there are indeed "many places in Scotland where the Erne could do no great harm and would be a glory to watch". Skye, I decided that shining morning, rather approved of my interpretation of the sea eagle's flight out over the sea from north Mull, for it had hung out the flags and unfurled the red carpet, the Cuillin glittered white (as they do in Duncan's painting), Beinn na Cailleach above Broadford wore a white skullcap, Broadford Bay was like crushed pale blue silk, and I was back in the landscape I have relied on more often than any other, more than a hundred visits in more than forty years. Skye does not push its eagles the way that Mull does, but like Mull, the potential to see one lies around every corner of the island, and my personal history here is replete with memorable encounters. I turned off the main road between the Skye Bridge and Broadford on a mission, for the grapevine had been active again. And speaking of places where "the Erne would be a glory to watch"…

⊙⊙⊙

The sea eagles at Glenelg have names and a website. Their names are Victor and Orla and the name of the website escapes me. I don't much care for giving birds names, nor for that matter websites that turn individual birds into celebrities. Despite the names and the website, I was bound for Kylerhea on a south-east shore of Skye, one of my favourite places anywhere these last forty years, and which, as it happens, has been part of the home territory of Victor and Orla since

2009, and I promise never to use their names again. The road to Kylerhea is a seven-mile cul-de-sac, unless you take the ferry from there to the mainland at Glenelg, in which case it is a passport to the rest of the world. It begins gently enough, but soon gets into its alpine stride, burrowing uphill into moorland and mountain with some daring. It is narrow, single-tracked, hairpinned, and on that astounding morning it was hell-bent on a high pass among frozen waterfalls. As if all that was not disconcerting enough, the views as you climb are in the west and over your shoulder, so you are forever consulting your door mirror where the Cuillin and Bla-Bheinn are constantly cropping up and realigning themselves, and time after time you are compelled to stop and point the camera at them then just stand and stare and marvel at the morning and your great good fortune that you resolved on a grey day on Mull to follow an eagle north.

Everything changes at the pass, where for a moment it looks as if you have run out of road, and for that matter, run out of land. The views are suddenly ahead of you, a startling sprawl of mainland mountains made more outrageous by the fact that someone has apparently removed a great scoop out of the foreground and that you are about to negotiate what is left of the mountainside. You are inclined to proceed by the thought that something special must lurk at the end of such a journey. It does.

If you have never witnessed at first hand a physical demonstration of the concept of irresistible force, go and sit on the rocks at Kylerhea and watch and listen to what happens at high tide when the Atlantic Ocean tries to cram itself into a rocky channel a quarter of a mile wide. It is a symphony scored by movement. If your idea of the ocean at high tide is the regular procession of waves, this will blow your mind, for there are no waves. Instead there is a high-speed charge through the narrows of water at its most ominously restless, given to sudden spasms in midstream, hissing loudly round the rocks at your feet as though it resented your presence. There are birds everywhere, mostly cormorants, gulls, ducks, divers, auks, goosanders, often going backwards at improbable speeds before they dive down into the maelstrom. An otter is bound to turn up sooner or later. (There is a public otter hide nearby, although I have never felt the need to use it and stillness on these rocks is all you really need.) Dolphins and occasional whales charge through here too.

The only quiet water is downstream of small headlands: in one of these a grey seal bull suddenly turned up, a piece of flotsam that suddenly acquired a head that snorted once, sank and disappeared.

Then the voice, high-pitched and zesty, an unearthly peal of wild *joie de vivre*, the song of the eagle. And at once in my mind I was out in an open boat in the deep temperate rainforest of south-east Alaska, cruising through inlets of the Pacific lined with steep mountainsides of spruce and hemlock, and a voice like that poured down from the trees and bounced across the flat surface of the water like a skimmed stone. At that moment, I had never heard such a thing. I turned to my host, a Tlingit Indian:

"What is *that*?"

"Oh, that's the eagle."

Oh, *that's* the eagle. And he showed me where it stood high in a tree just above the shore. And after that, once I'd got my eye in, there were the white heads and white tails of bald eagles in tall trees every half-mile or so. The south-east of Alaska teems with bald eagles and that voice is the anthem of that land. And now something very like it has returned to a Skye shore which also once claimed an eagle voice for its anthem, not the occasional terrier yap of the golden eagle but the much more widespread carillon-like signature of the sea eagle. Robert Gray's 1871 book, *The Birds of the West of Scotland*, noted that "Skye was the headquarters of this conspicuous eagle nearly all the bold headlands of Skye are frequented by at least one pair..."

The eagle-killers of the day agreed: a nineteenth century landlord, one Captain Cameron of Glen Brittle, conceded that of sixty-five eagles he either killed "or caused to be killed", only three were golden eagles. And Gray notes that one keeper shot fifty-seven in nine years on one Skye estate. No surprise that by 1904, the newly published *A Vertebrate Fauna of Scotland* recorded that there were only "a pair or two" breeding on Skye. And after 1916 there was nothing at all on Skye until 1988.

So that voice spilled across the Kylerhea rocks from somewhere above and behind me, and I said to myself, "Oh, that's the eagle." And for a moment I bestrode two lands five thousand miles apart and the thing that united them was the brotherhood of eagles. I eased myself slowly up from my seat among the rocks to see if I could pin down the source of the voice. With and without the binoculars I could find nothing at all. I contented myself for the moment with the knowledge that there was at least one bird in the area, and the chances are that it was calling to another one. So I settled again and waited, and watched the sea, the birds that paraded up and down the Sound in front of me, and all the while the boxy little car ferry Glen Achulish pursued unlikely (to this landlubber) courses back and forward through the madly furrowed water, to the perpetual accompaniment of the skipper's collies, which are a compulsory part of the process. If I didn't do this for a living (I do), and if I was fluent in the speech of colliding tides and currents (I am not), I could think of many worse jobs than a ferryman at Kylerhea. As it is, I have watched and admired from a distance for all these years and occasionally contributed to its welfare as a fare-paying passenger, eschewing the cumbersome intrusion of the Skye Bridge for the privilege of patronising the oldest of all Skye ferry crossings. Now with added sea eagles.

I heard two crows across the sound, and looked up. Crows are mostly just talking to other crows when you hear them, but sometimes they are scolding the presence of something they disapprove of in the neighbourhood, so in a place like Kylerhea, when I hear crows I look up. They were quite far off but the calling went on and on, so that I started to take them seriously and began to scour the low ground on the far shore with the binoculars.

I found nothing there so I slowly worked my way higher and from side to side and I found six crows behaving madly in short roller-coaster flights, each of which bottomed out near two very large birds moving slowly in wide sun-wise circles. On such a beautifully lit day and even at that distance, their white tails shone in the sun. They climbed and they climbed and they climbed, always in slow circles, and they were still climbing when they left the trees and the crows behind. They were then against difficult hillside, but soon they would be against snow and then blue sky if they kept going. I scanned the glasses vertically down to the shore to pick a marker there in case I lost them against that difficult background, and was about to raise them again when an otter sprang into sharp focus in midstream.

It had powered up out of the very fastest part of the current, half out of the water, jaws open, and then it dived under again. What now? Back to the eagles or wait for the otter to resurface, and what if it caught a big something or other and brought it ashore, the way otters do? Would I lose the eagles I had come here to see? I waited a few seconds more for the otter, then went back to the marker on the far shore and raised the glasses from there to find the eagles, hoping against hope that they had not decided to lose themselves against the trees again en route back to the nest. But they were still climbing, and then they were against the mountain snow and the sky, and still they climbed, but they started to drift south as they climbed and now my marker was redundant. Stay with them, I counselled myself, for as long as it takes for something to happen, or until my arms get tired, which, as it turned out, was the only thing that happened. They climbed until they looked like motes of dust on the lenses, and I gave in, I surrendered to their eagle-ness. The otter, of course, was not in sight.

I had a quiet lunch on the shore, and I thought about Leslie Brown again. His book reveals a deep thinker on what were the unanswered eagle questions at the time. One of those questions is partly the purpose of this book.

"I am most intrigued at present by the mystery of what happens to the eaglet after he leaves his parents, for this is the part of an eagle's life that is difficult to follow." What he would have made of satellite tracking and wing-tagging, and reintroductions using two-month-old chicks rather than eggs, and raising them with human rather than golden eagle foster parents, is difficult to imagine. But there, soaring beyond snow-scrolled mountains on Scotland's Hebridean shore, are the fruits of your labours, Leslie, and the answers to the questions you posed for new generations to grapple with. And oh, the glory of the watching!

Chapter 11

RETURN TO
THE EAGLE GLEN

*The golden eagle repays prolonged observation
more than most birds.*
– Seton Gordon, *The Golden Eagle* (Collins, 1955)

THROUGHOUT THE RESEARCH and writing of this
book, and for thirty years before I ever thought of writing an
eagle book, my endeavour had been to establish myself as a
predictable, benevolent presence in the glen that has always
accommodated the nearest eagles to my home. Year after year
and season after season, I became a mobile fragment of the
chosen landscape of eagle after eagle. It may not have proved
the most fecund of golden eagle territories but nor has it ever
lacked adult eagles, and these have provided much of my
schooling and my graduation; the big rock that faces the eyrie
buttress from the distant floor of the glen and another on the
watershed that commands a view over almost the whole territory of the resident pair – these have been my seats of learning. This is where I began to understand about the nature of
golden eagle territory, and where I was taught by the eagles'

own example about the wisdom of establishing a home territory of my own as a nature writer. So I worked the same landscape circumstances again and again, season after season and year after year until something like intimacy with my surroundings was within reach. That intimacy now informs almost everything that I write, for once I had acquired at least a working knowledge of the principles and patterns of nature within a specific landscape that embraces elements of both Highlands and Lowlands, I found that these are transferable, that they also hold good wherever I have travelled. This then is the golden eagle territory that taught me how to work; these then (the pale female and her smaller, darker mate) are my workaday eagles.

It has taken time. When I first got to know about the glen I was still a newspaper journalist and a spare-time mountaineer, and the idea of becoming a nature writer was a sometimes-dream. The eagles' fortunes were at a low ebb thanks to persistent and ingenious egg thieves. Almost by accident I became part of an organised watch through the nesting season. Eventually the watch prevailed and young eagles began to fly free from that glen again, at least as often as not. Meanwhile, I had begun to see eagles often and to become slowly and irrevocably spellbound. When my first season with the eagle watch came to an end I suffered withdrawal symptoms, until it occurred to me that there was nothing to stop me from watching on my own, now that I knew where they lived. Or at least, there was nothing to stop me from walking the hills of their territory in the hope of meeting them out in their wide world away from the confines of the eyrie. And so it happened. There were blank days of course, but golden days too.

I started to watch eagles being eagles. These were no longer the chance encounters that every mountaineer knows, but rather they were the direct consequence of the fact that I was looking for them with eyes wide open and a new thrill in the back of my throat. In the course of two or three years I lost interest in the idea of climbing a mountain to reach the top. Instead I was lured by eagle-thraldom into a different relationship, first with the mountain world then with all wild landscape, a relationship whose objectives became the unravelling of nature's secrets, a better understanding of the wild world.

In a way it was like a reversion to childhood, for this was how it all began for me, not with mountains and eagles but with skylarks and swans, hares and hedgehogs, curlews and kittiwakes, all along the Tay estuary and deep into its fringing fields and woods and low hills. This new way of moving *among* mountains was my old way, my first way of meeting nature head-on. It was instinctive and true, and it fitted me like a glove. Then, of course, I decided I would like to try and write it all down, and twenty-six books later, here is my first eagle book.

The eagle glen is not well disposed, it seems, to promote longevity in its eagles. In more than thirty years year I have seen several "new" birds take over the territory without always knowing what fate befell their predecessors. As I write, the current female, the pale beauty from the Prologue, has been here five or six years, during which time she has raised not one chick to fledging. Her immediate predecessor was huge and dark, and she had occasional successes, but her only two-chick year was a prelude to a particularly vicious turn of fate. Both chicks fledged and left the territory in the winter, but one was found dead in a crow trap and the other was poisoned. I did not see the dark female again either, and I wondered then as I wonder now, if there was a grim conspiracy to rid the glen of eagles that year. No one should be in any doubt that *some* landowners, *some* gamekeepers, *some* land managers are relentless in their pursuit of everything in nature they judge inconvenient for the way they run estates. Grouse moors and deer forests are as unnatural landscapes as a forest of dense Sitka spruce, and just as hostile to nature as any kind of monoculture. Shooting, trapping and poisoning are routine, and if their perceived problem happens to be a golden eagle then the highest level of protection the law affords such birds will not save them, any more than it will save a fox or a stoat or a raven or a crow. And now that the sea eagle is re-establishing itself, now that it has shown a propensity to cross the country from coast to coast, now that its population is growing at a pace that will inevitably lead to the day when it outnumbers the golden eagle, as it did throughout history until we took a hand... now the sea eagle is also right at the top of the hit list. The worst excesses of Scottish landowning practice will begin to change when the law is willing to stand up to them and jail a landowner, as well as the culpable keeper who is caught doing his bidding.

A hundred years after the sea eagle was wiped from the face of the land, the same attitudes that achieved that outcome are resurfacing: the first poisonings, the first destruction of an east coast nest, and the farming industry has proved itself to be quite capable of "it'll be our children next" bleatings when a lamb is found inexplicably dead in a field that a sea eagle may or may not have crossed. Imagine the cacophonous outpourings when the sea eagle population reaches five hundred pairs.

Meanwhile, government enthusiasm for the wind farm industry and landowners' willingness to see their lands wrecked and turned sterile by it, has turned the screw that little bit tighter in the tricky balancing act of sustaining Scotland's eagle population. Habitat is everything, and an industry that destroys it and kills eagles as a by-product is a barely creditable artefact of twenty-first century civilisation. As a country we are no better at accepting a conspicuous new presence of nature in our midst than we are at according established conspicuous presences a right to co-exist; we are reluctant to admit to the Victorian time-warp that still strangles even the most enlightened of conservation endeavours and still finds ways on dark nights and isolated places to outwit new laws aimed at wildlife protection. Too much land is still in private hands, too much power concentrated in too few landowners whose motivation for owning the land is rarely the wellbeing of the land, and almost never to provide a better place for nature. Many overseas visitors to our national parks are dumbfounded to learn that they are not owned by the nation but rather by fractious coalitions of private individuals. Eagles do not fare well in our national parks, they find little enough sanctuary there.

All this was the inheritance of the pale female in the eagle glen when she inadvertently became the heartbeat of my own adopted territory. It was April in her sixth year, and by then I had serious doubts about the viability of her eyrie. A bush on the eyrie ledge had mushroomed over the past two years and appeared to me to obstruct the birds' access. Once that month I saw the male fly in, vanish behind the bush then reappear almost at once. I thought that it had become too difficult for birds with a wingspan around two metres. At the very least, the sitting bird would have no view of the glen, and see no more than a wedge of sky. Is that a problem? I don't know, and there is probably no definitive answer. Such is the individuality among eagles that it might not faze one bird at all but be intolerable to another. If those differences co-existed within a pair, then yes, there is a problem. Subsequent but spasmodic visits convinced me that the birds were not using that eyrie. All golden eagle territories hold alternative nesting sites but there is almost always a favourite, and in more than thirty years an alternative to the eyrie on the buttress was only used once to my knowledge.

Spring was long and wet and cold, as the winter had been, conditions that can cause golden eagles to simply abandon nesting for the year. Then there was the long hot summer that surprised everyone, and nature made haste and seemed to catch up with itself and many of the glen's wild tribes responded to the warmth and fulfilled the year's destiny. But too often there were two eagles in the air for too long when (it seemed to me) there should have been one on the eyrie, and too often there were no eagles to be seen. The portents were not well aligned. I finally accepted the worst over two long hot July days when I scoured the glen, the watershed and as many square miles of the territory as I could handle, and saw nothing at all. But the particular problem with such days in such an uncharacteristic year is that even the most experienced and thoughtful of eagle-watchers might be forgiven for responding to them with a conclusion rather than a question. I did just that. I was wrong.

It was over a month later before I was back in the glen, with a question. In the interim I had made a slow, two-way sweep of my notional eagle highway from coast to coast and one last trip to Skye. I had seen eagles everywhere I had expected to find them. Back home, I was uneasy and restless. I decided I owed it to the glen that had taught me so much to try and improve my understanding of the eagles' life there. The question I took with me was this:

"What am I missing?"

That question fostered others: what changed either on the eyrie ledge or in the nature of the territory; did the eagles try to nest somewhere else and did that explain their failure; did that also explain why I had seen them so little of them, because the glen's airspace was no longer part of the land-scape of the new eyrie? All along the eagle highway I had alighted on eagle places whose nuances were new to me, but I expected more from this glen, not least because I had given it so much more and because I thought I knew the ropes. And all I knew for sure when I came back that August day was that I was missing something. And of course, I saw no eagles. Fair enough, I told myself, it happens. On my next visit a few days later I saw no eagles and left the glen shadowed once more by old forebodings. I went back again with the same result, despite putting in eight hours out on the territory.

On the fourth trip I met a party of visiting birdwatchers clustered at the end of the forest road around telescopes and cameras with big lenses on tripods, and in the care of a local man who runs wildlife safari trips from a big Land Rover. He said he was having a good season, that they had been seeing eagles every other trip on average, but at no time had he seen a young bird. He wondered if there was enough feeding for the eagles, what with the catastrophic fall in mountain hare numbers, and as sheep farming had almost become extinct in that country. Then he said that earlier on there had been a gap when he had seen no eagles for six weeks.

I left the group and walked up into the high part of the glen, which in late summer has the scent and feel and look and sheen of an alpine meadow, and beyond the last of the plantation trees to where there are prospering mountain birchwoods. I followed the burn, pausing often to watch the sky, the crags, the ridges on three sides, and all the middle distance. I took in the flowers, the scurries of small fish, the voices of pipits and late warblers still singing, all things bright and beautiful that might deflect my mind from gloomy thoughts. It was not a day or a place for gloominess, yet "no eagles for six weeks…" And what might *that* signify? Something of Seton Gordon's slipped into my mind as it often does when I am wrestling with eagle nuances:

"Perhaps as the years pass they are no longer inseparable during the winter months."

And might that hold good too for high summer if it had been yet another barren nesting season and the pair's sixth successive failure, their wild year robbed of all meaning yet again? Might that also have the effect of loosening the pair bond so that they drifted alone beyond the territory?

The day grew warm in mid-afternoon and I settled high up for a late lunch and a long stillness. A pair of ravens grew loudly agitated, so that I snatched up the binoculars, but it was the hanging persistence of a kestrel that was bothering them. Ravens know almost every trick of flight in the book including the upside-down trick, every trick but not the kestrel's deadstill hover, and this one was leaning on thermals rather than wind, for there was no wind, and in the bird's airy stance there were no moving parts, and I told myself the ravens were miffed at the kestrel's accomplishment that was beyond them. Ravens make me smile, but I have an irrational fondness for the kestrel.

The hours drifted and so did my concentration – it was time to go home. I stashed my empty cup and sandwich box and folding seat mat, camera, notebook and pens, and as I bent to close the rucksack a new movement lodged in the corner of my right eye. I straightened and turned to face the watershed and there were three golden eagles in my sky.

☻☻☻

Three eagles? How on earth could there be *three* eagles? They had just crossed the watershed from the north when I turned and saw them, widely spaced and dark against a sky of electric summer blue patched with towers and mythic palaces of cumulus white. If I believed that such a thing was biologically possible, I would say that my heart lurched. But for sure, something fundamental realigned somewhere inside me. That was how it felt at that moment. Three eagles, and oh, the glory of the watching!

They came together above the ridge and began to travel towards the buttress. Then just before they reached it, they slipped one by one down from the ridge and into the glen's airspace, and they flew straight across the front of the eyrie ledge, passing within a few yards of it. I heard quite clearly the shrill terrier yap of the young bird. They held formation for another quarter of a mile then rose back into the sunlight, crossed the ridge again and disappeared, the ragged-edged eaglet and svelte dark male in line astern of a conspicuously larger female whose plumage glowed as pale as desert sand. They had crossed low over the ridge, showed only briefly against the sky as the blackest of silhouettes, the eaglet's wings working without fluency compared to the wide-winged glide of its parents. Then they had tilted south-west and were gone and I stood and stared at the newly-emptied sky, and my mind had gone numb and I was briefly beyond rational thought.

I began to walk, retracing the day's trek back down through the glen and the cooling forest beyond and I was bewitched, bothered, bewildered and elated, and not making much sense of anything at all. I could not believe – I still cannot believe – that the eaglet had flown from the old first-choice eyrie, and yet there had been that deliberate diversion down from the ridge and the flypast right in front of the bush that obscured the eyrie; so what was that all about if it was not to confirm the family's place on the map of the glen?

I reached my car in the same tumultuous state of mind. I stopped on the drive home and found a seat in a quiet bar, sat there for an hour with a single beer and an open note-book and when I left the beer was half-drunk and warm and the notebook page was empty. I had thought I was missing something, I was right, but I am still no nearer to understand-ing what happened or to know precisely what I was missing. But a new eagle had flown from a barren nest and it felt as if something precious beyond price had been restored to me. And then, a week later, I heard on the radio that the RSPB had just confirmed that a sea eagle chick had fledged from an eyrie in a forest in Fife, the first on the east coast for two hundred years.

EPILOGUE:
WHAT NEXT?

SO I REACHED OUT effortlessly for symbolism. Why would I not? A golden eagle hatched and fledged and flew unseen and unsuspected in the glen where looking for eagles has been a routine for half my life. And a sea eagle hatched and fledged and flew from an east coast woodland deep within the force-field of the Tay estuary, the first of all my land-scapes. Both of these flights happened the same week that I had written *finis* at the end of this book's manuscript. There are times in the nature writing life when nature appears to tap you on the shoulder, and you do well to pay attention.

The young golden eagle is the first to have emerged from the womb of that glen after seven barren years, and the first offspring of the beautiful pale eagle which has become my mind's eye's definition of all her tribe, despite all the eagles that have crossed my path from Sutherland and St Kilda to the hills of home and the hills of Galloway. The young sea eagle is the first east coast native in two hundred years and, given a fair wind, the harbinger of an east coast settlement, a bird loaded with significance and omen, for although more than eighty birds have been brought from Norway to fly free over east coast skies the enterprise is without substance until native birds establish themselves and evolve their own way of life into which more and more generations are born. If the enterprise had failed, if all the Norwegian-born east coast arrivals had ultimately tended westwards, the eagle highway would also surely have failed for want of new blood from the east. A few birds from the west might still have occupied the historic inland haunts between Loch Awe and Rannoch, but the impetus to travel east beyond the Highlands would not survive without the pull of an east coast population. All the ecological riches that are bound to flow back and forth from coast-to-coast travel and settlement would be lost – riches we can only guess at yet, in much the same way that no one anticipated the beautifully benevolent extent of wolf reintroduction into Yellowstone when that enterprise began. The scale may be different, and Scotland is, alas, still some distance away from putting wolves back into Rannoch, but the principle of repairing a systematically raided ecosystem by installing new native blood at the top of the food chain is one that cannot fail. That, and the restoration and expansion of every native habitat is all the help nature needs to recreate something of an older, wilder order.

The first thing we will learn is that the golden eagle will not be impoverished if and when – almost certainly "when" – it is outnumbered by the sea eagle. For that is Scotland's historical norm, and it is the situation in Norway today where there are several thousand pairs of sea eagles and a much smaller but thriving population of golden eagles. And mostly, the two eagle tribes co-exist amicably enough. In occasional one-to-one skirmishes, usually a territorial dispute, I have never seen and never heard of a sea eagle prevailing. The

golden eagle is the supreme flier as well as the supreme predator of our skies, and that alone should be enough to guarantee the stability of Scotland's golden eagle presence. Inferior
numbers will not change that.

A wise man, a grizzly bear guide on Kodiak Island,
Alaska, once told me: "The only way to get to know a creature is to live with it." We were speaking in his cabin on a
tiny island in the middle of a lake on Kodiak where I had
found moulted bald eagle feathers by his doorstep, and the
subject under discussion was living with bears, with wolves,
with eagles. He was experienced and adept at sharing the
landscape with all of these. He was the one who convinced
me by his own example that we learn about and understand
other creatures only by living with them, not by taking the
word of the loud mouths and small minds of those vested
interests who would shout us down and have us believe that
Scotland cannot accommodate those creatures their predecessors cleared from the land. Scotland can. The land can
and the people can. We learn by living with them. We make
adjustments and so do they.

Most of us think of ourselves as living apart from nature,
but the other creatures of nature think of us as a part of
nature. They see us as powerful and unpredictable, sometimes lethal and sometimes generous, a serious presence in
the land and at times a formidable predator. So where they
live close to us, or when we take a long walk to where they
live, they make adjustments to our presence. That is the way
it has always been. New wolves, new bears, new eagles will
behave the same way. That is how it should be.

If the sea eagle really takes hold in the east, many more
people will start to see them regularly, the people and the

eagles will be fixtures in each other's landscape. We learn about them by living with them. They learn about us by living with us. In the same way, the two eagle tribes learned to live that way with each other many, many thousands of years ago.

So what's next? I honestly don't know. And neither does anyone else. What I do know is that I want to be there when it happens. I can still hardly believe my luck at what has begun to evolve before my nature writer's eyes, for the eagle's way just happens to be my way too. Whatever unfolds, it will never be less than instructive, never less than spectacular, never less than wild. If I were you, I'd watch. And oh, the glory of the watching!

ACKNOWLEDGEMENTS

I am grateful to Creative Scotland for a grant that greatly assisted this book's cause, especially in funding almost constant travel throughout the research and writing. I thank them too for their professional assistance, which was never less than helpful and encouraging.

Warm thanks to Jenny Brown, my agent, who found the book a good home, and to Sara Hunt, Craig Hillsley and all at Saraband who have provided that good home.

I am especially pleased that this book features the photographs of Laurie Campbell, who is simply among the very best in the business. We have talked off and on for years about the possibility of collaborating; this is a particularly appropriate project in which to make it happen.

To the friends who contributed their eagle stories – Polly Pullar, Ann Lolley, Mike Holliday and Duncan Currie – thanks so much for responding so enthusiastically, and the drinks are on me.

I have benefited from many conversations over many years with eagle enthusiasts, too many of whom are no longer with us. In particular, these include dear friends Andrew Currie who died on Skye while I was writing this book, and Pat Sandeman. Their friendship is affectionately acknowledged as readily as it is missed.

This book is a drop in the ocean of eagle literature of which there is no more towering presence in a Scottish context than Seton Gordon; his books continue to inspire and challenge me, his legacy to Scottish nature writing is immeasurable. His *Days with the Golden Eagle* and *The Golden Eagle* are landmarks of Scottish nature writing.

Other writers whose work I have summoned to my cause include Leslie Brown, whose book *Eagles* offered a glimpse of just how unlikely the return of the sea eagle appeared even among eagle experts as recently as sixty years ago, and Hugh Johnson, whose monumental book *Trees* is among the most thumbed in my bookshelves. The others are friends who one way or another have enriched my life and whose work I quote as much in gratitude for that friendship as for the particular way it serves the cause of this book: George Mackay Brown for his poem *Bird and Island*; Marion Campbell of Kilberry, whose *Argyll – The Enduring Heartland* is simply one of the finest books anyone ever wrote about Scotland; and Mike Tomkies, whose *A Last Wild Place* gets under the skin of the wild Highlands in ways that few other writers have ever dared. My thanks to all of those writers go far beyond the fact that they dignify the pages of this book.

FURTHER READING

BROWN, LESLIE, *Eagles*, Michael Joseph, 1955.

GORDON, SETON, *Days with the Golden Eagle*, Williams and Norgate, 1927, and Whittles Publishing, 2003.

GORDON, SETON, *The Golden Eagle*, Collins, 1955, and Melven Press, 1980.

LOVE, JOHN, *The Return of the Sea Eagle*, Cambridge University Press, 1983.

LOVE, JOHN, *A Saga of Sea Eagles*, Whittles, 2013.

TOMKIES, MIKE, *Golden Eagle Years*, Jonathan Cape, 1982.

WATSON, JEFF, *The Golden Eagle*, Poyser, 1997.

ABOUT THE AUTHOR

Jim Crumley has written more than twenty books, many of them on the wildlife and wild landscape of his native Scotland. He is a widely published journalist with regular columns in *The Courier* and *The Scots Magazine*, a poet, and occasional broadcaster on both radio and television.